VLSI SIGNAL PROCESSING TECHNOLOGY

VLSI SIGNAL PROCESSING TECHNOLOGY

edited by

Magdy A. Bayoumi

University of Southwestern Louisiana

and

Earl E. Swartzlander, Jr.

University of Texas

KLUWER ACADEMIC PUBLISHERS
Boston / Dordrecht / London

Distributors for North America:
Kluwer Academic Publishers
101 Philip Drive
Assinippi Park
Norwell, Massachusetts 02061 USA

Distributors for all other countries:
Kluwer Academic Publishers Group
Distribution Centre
Post Office Box 322
3300 AH Dordrecht, THE NETHERLANDS

Library of Congress Cataloging-in-Publication Data

VLSI signal processing technology / [edited by] Magdy A. Bayoumi and
  Earl E. Swartzlander, Jr.
    p. cm.
  Includes bibliographical references and index.
  ISBN 0-7923-9490-9 (alk. paper)
  1. Integrated circuits--Very large scale integration. 2. Signal
processing--Digital techniques. I. Bayoumi, Magdy A.
II. Swartzlander, Earl E.
TK5102.9.V54 1994
621.382'2--dc20   94-3662
                                      CIP

Printed on acid-free paper.

Printed in the United States of America

To My Dear Students,

I live your worries and ambitions as a father.

I live your ups and downs, smiles and tears, courage and fears as a brother.

I live your adventurous and mysterious journey as a friend.

You have made my life spiritually rich, intellectually rewarding, culturally colorful and heartly young.

I'll never forget you.

Magdy Bayoumi

TABLE OF CONTENTS

LIST OF CONTRIBUTORS

Magdy A. Bayoumi
The Center for Advanced Computer Studies
P.O. Box 44330
University of Southwestern Louisiana
Lafayette, Louisiana 70504

Vasudev Bhaskaran
Hewlett-Packard Laboratories

C. T. Chiu
Electrical Engineering Department
National Chung Cheng University
Chia-Yi, Taiwan

Eby G. Friedman
Department of Electrical Engineering
University of Rochester
Rochester, New York 14627

Yoshihiko Horio
Department of Electronic Engineering
Tokyo Denki University 2–2
Kanda-Nishiki-cho
Chiyoda-ku, Tokyo, 101, Japan

G. A. Jullien, Director
VLSI Research Group
University of Windsor
Windsor, Ontario, Canada N9B 3P4

Konstantinos Konstantinides
Hewlett-Packard Laboratories

K. J. Ray Liu
Electrical Engineering Department
University of Maryland
College Park, Maryland 20740

J. H. Mulligan, Jr.
Department of Electrical and Computer Engineering
University of California
Irvine, California 92717

Shogo Nakamura
Department of Electronic Engineering
Tokyo Denki University 2–2
Kanda-Nishiki-cho
Chiyoda-ku, Tokyo, 101, Japan

Ross Smith
University of Illinois at Chicago
Department of Electrical Engineering
and Computer Science (M/C 154)
Chicago, Illinois 60607

Gerald Sobelman
University of Minnesota
Department of Electrical Engineering
Minneapolis, Minnesota 55455

Mani Soma
Design, Test & Reliability Laboratory
Department of Electrical Engineering, FT-10
University of Washington
Seattle, Washington 98195

Hiroyuki Takase
Computer Works
Mitsubishi Electric Co. 325
Kamimachiya
Kamakura, Kanawaga, 247, Japan

Preface

This book is the first in a set of forthcoming books focussed on state-of-the-art development in the VLSI Signal Processing area. It is a response to the tremendous research activities taking place in that field. These activities have been driven by two factors: the dramatic increase in demand for high speed signal processing, especially in consumer electronics, and the evolving microelectronic technologies. The available technology has always been one of the main factors in determining algorithms, architectures, and design strategies to be followed. With every new technology, signal processing systems go through many changes in concepts, design methods, and implementation.

The goal of this book is to introduce the reader to the main features of VLSI Signal Processing and the ongoing developments in this area. The focus of this book is on:

- **Current developments in Digital Signal Processing (DSP) processors and architectures** – several examples and case studies of existing DSP chips are discussed in *Chapter* 1.

- **Features and requirements of image and video signal processing architectures** – both applications specific integrated circuits (ASICs) and programmable image processors are studied in *Chapter* 2.

- **New market areas for signal processing** – especially in consumer electronics such as multimedia, teleconferencing, and movie on demand.

- **Impact of arithmetic circuitry on the performance of DSP processors** – several topics are discussed in *Chapter* 3 such as: number representation, arithmetic algorithms and circuits, and implementation.

- **Pipelining** – which has greatly effected the performance of DSP systems. Several factors are involved in designing pipelined architectures. They are: the trade-off between clock frequency and latency, clock distribution network, and clock skew as shown in *Chapter* 4.

- **Parallelism** – to meet the high computational rates of real-time signal processing, inherent parallelism of most signal processing algorithms has to be exploited. To achieve the required performance, it necessitates developing special-purpose architectures featuring regularity, modularity, and locality as shown in *Chapter* 5.

- **Systolic Arrays** – are one of the most efficient parallel architectures for signal processing. A detailed design of a systolic cell along with the controller is discussed in *Chapter* 6.
- **Analog Signal Processing (ASP)** – The main factors for the recent resurgence of interest in ASP are: the recent interest in neural networks circuits and applications, and the advances in ASP technology which permits better control of device characteristics. In *Chapter* 7, four major analog circuits design are discussed: continuous-time, switched-capacitor, switched-current, and subthreshold analog circuits. In *Chapter* 8, a case study of an application specific analog architecture for speech recognition is discussed. It is based on parallel distributed processing network of switched capacitors.

The intent of this book is to be informative and to stimulate the reader to participate in the fast growing field of VLSI Signal Processing. The book can be used for research courses in VLSI Design and VLSI Signal Processing. It can also be used for graduate courses in Signal Processing and VLSI Architectures. The chapters of the book can serve as material for tutorials and short courses in relevant topics.

Finally, we would like to thank the authors who spent considerable time and effort to have their research work reported in this book. We appreciate their patience through the whole project.

Magdy Bayoumi
Earl Swartzlander

VLSI SIGNAL PROCESSING TECHNOLOGY

1

VLSI DSP Technology : Current Developments

Magdy A. Bayoumi

The Center for Advanced Computer Studies
P.O. Box 44330
University of Southwestern Louisiana
Lafayette, LA 70504

1. INTRODUCTION

The field of digital signal processing has received a great deal of impetus, over the past decade, from the technological advances in integrated circuits design and fabrication. During the 1970's, the hardware available was mostly in the form of discrete TTL (transistor-transistor logic) or ECL (emitter-coupled logic) devices. Although a reasonably high level of performance was available from such implementations, the hardware design was so massive and expensive that most often unsatisfactory hardware/performance trade-offs were the order of the day. With the advent of high levels of integration on a single silicon substrate, new applications were introduced, particularly at the consumer level. Thus, compact disc players, that use reasonably sophisticated DSP techniques, are available at a very low cost to the average person. In turn, this hardware revolution reflects back on the research community which responds with new directions that can take advantage of the technology. It is the aim of this chapter to briefly review the current developments of VLSI DSP technologies and their applications.

2. VLSI DSP TECHNOLOGY

Currently, DSP technology is where microprocessor technology was in the early 1970's. It is expected that an explosive growth will take place with fast advances in DSP applications. The confluence of faster and denser silicon with the increasing sophistication of DSP algorithms has resulted in many applications previously handled by board level systems, now being taken over by single-chip solutions. In additions, as these single chips advance along the cost/volume learning curve, they begin to yield cost effective DSP solutions to problems that have traditionally and until now, most economically been handled using analog technology. The new generation of CAD tools, system software, and application tools, are emerging, too.

DSP architectures have distinct features compared to general computing systems due to the special characteristics of DSP algorithms. The main criteria of DSP algorithms are:

1. They involve intensive amount of data.
2. They require multiple use of the same data.
3. There may be feed back paths (e.g. IIR filters).
4. Fast multiplications are required.
5. Intensive intermediate data manipulation and address generation.
6. Most of the algorithms are amenable for parallel computation.

In view of the above requirements and speedup of computation, the main features of DSP architectures can be summarized as follows [8,13]:

1. **Fast arithmetic units**: High speed adders and multipliers can be realized at both technological and circuit levels. As an example, varieties of adders such as carry-look-ahead adders, carry-save-adders, carry-select-adders, etc. are generally proposed to meet the fast computational requirements.

2. **Multiple Functional Units**: Since operations in DSP are vector or matrix operations, most of them can be performed either in parallel or a pipelined fashion. This requires multiple functional units.

3. **Parallel Computation**: It is important to identify the parallelism and decompose the given algorithm into dependent tasks.

4. **Pipelined Functional Units**: Pipelined computation leads to high throughput rates.

5. **Proximity of storage**: This is probably one of the predominant factors influencing the speedup in performance. Most of the algorithms require the accessing of constant weights, stored as values in special ROMs. Consequently, the access rates of the constants weights should not become a bottleneck in the performance of the computation. Hence the stored values might be in a distributed fashion rather than having a global memory or ROM.

6. **Localized Data Communication**: This requirement is a natural consequence of achieving high speed pipelining. Limiting the communication to local modules helps in achieving high speeds.

7. **High Bandwidth Buses**: Since date transactions involve transfers large amount of data, the design should be aimed at reducing the total time involved in these data transactions. This can be achieved with high bandwidth data buses.

VLSI DSP architectures can be classified into three main categories: *general purpose chips, core-based systems, and application specific chips.* General purpose DSP chips is the fastest growing class. Texas Instruments has been the leader in the area with its 320 family, a variety of NMOS and CMOS devices. The main developments have been taking place in the floating point chips, parallel architecture chips, and RISC DSP chips. Floating-point devices provide a far greater dynamic range, simpler programming and greater resolution than the fixed point counterpart, they can offer significant savings in the up-front programming and low-volume assembly costs. AT&T produced the first commercial 32–bit floating-point chip in Fall 1986. The performance of these chips can be enhanced considerably by including high degree of parallelism in performing the involved arithmetic and I/O data transaction. While DSP chips are generally considered special RISC processors, a new generation of RISC DSP chips has evolved. They include the main feature of RISC architectures. Intel 80860 has been a very good example for this category. The DSP core approach is a compromise between the general purpose DSP chips and the dedicated systems. It combines the cost-effectiveness and ease of programming of the first, and the high performance and flexibility of the later. DSP controllers are the most common configuration in this category. The third category is the application specific DSP chips which are being developed for dedicated applications where the required performance cannot be met by the other approaches.

3. GENERAL PURPOSE DSP CHIPS

By far, the fastest growing segment of the DSP processors market is the general-purpose chips. Texas Instruments has been the leader in this area. These chips are characterized by the presence of multiple functional units and the basic components of these chips are the multiplier, data RAM, fast ALU, and a program memory sufficient to store all the data

4

on the chip. These also contain multiple buses and a set of high speed
registers. A generalized structure of a general purpose DSP chip is shown
in Fig. 1 [1]. A key element on a general purpose DSP chip is a fast
array multiplier in a single clock cycle. A typical fixed-point multiplier-

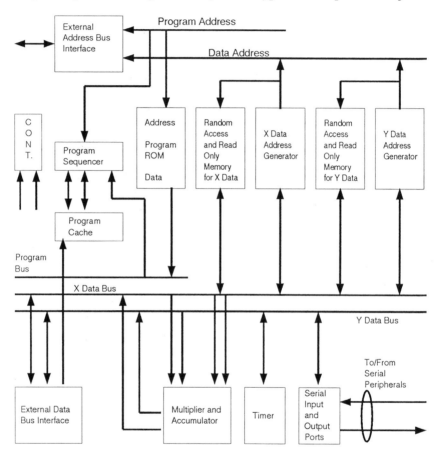

Figure 1 Architecture of General Purpose Digital Signal Processors

accumulator performs a 16 by 16 multiplication and the 32 bit product is
added to a 32 bit accumulator register in a single instruction cycle. An
alternative design uses separate elements for multiplication and addition,
as in μpd7720A and μpd77230 digital signal processors [2], made by
NEC Electronics Inc. This approach allows the system engineer to fine
tune the processor for specific application by controlling the software
pipeline of the multiplication and addition operations.

Multiple buses and multiple memories make DSP chips different from the other processors like microprocessors. A typical general purpose digital signal processor has two data memories and two data buses. We can deliver two operands required for single-cycle execution of a multiply accumulate function. Unlike the general purpose microprocessors, which store both the instructions and the data in the same memory, most DSP processors employ the Harvard architecture [2], with separate program and data memories so that both the instructions and data can be fetched simultaneously. Many DSP chips go a step further, employing a modified Harvard architecture to allow storage of data in the program memory. In this way, static data like filter coefficients can be stored in a cheaper and slower program memory and then be moved to faster but smaller data memory when needed.

There are a variety of ways to enhance the system's throughput by increasing the rate of data access (memory bandwidth). This is usually accomplished through multiple memories and corresponding buses. Several DSP chips have dual internal data memories as well as access to an external data memory. Typically internal memories contain 128 to 512, 16 bit words, adequate for many DSP tasks. Other DSP circuits have mechanisms for fetching instructions from internal program RAM, there by freeing the external bus for additional data access. In the ADSP2100 chip, from Analog Devices Inc., this is achieved with a 16 word program cache that allows two external data access in a single cycle. Another DSP chip, the TMS320C25 from Texas Instruments[2], has a repeat facility, by which a single instruction can be repeated a specific number of times with only one single fetch from the external program memory. Again, when the code is running internally, an external data bus which is otherwise dedicated to accessing external code is free to access data. The design of TMS320C30 from Texas Instruments calls for a 64–word program cache. It also includes direct memory access arrangement on the integrated circuit, which transfers blocks of data from an external memory into DSP chip's arithmetic and logical unit. Texas Instruments TMS32010/32020 signal processor was designed with high speed digital controller with numerical stability [3]. The architecture of the chip is given by Fig. 2. Program and data buses are separated, which allows full overlap of instruction fetch and execution. TMS32010 was fabricated in 2.7 micron NMOS and packaged in 40 pin dual-in-line package. It has 1536x16 data RAM. The arithmetic unit has 16x16 bit array multiplier.

6

Shifters are provided for data shifting.

Fujitsu MB8764 signal processor [3] shown in Fig. 3 is a very fast CMOS monolithic signal processor. It consists of seventeen functional units, independent clock generator, sequencer, addressing section, RAM, and ALU which has multiplier and I/O sections. The instruction exe-

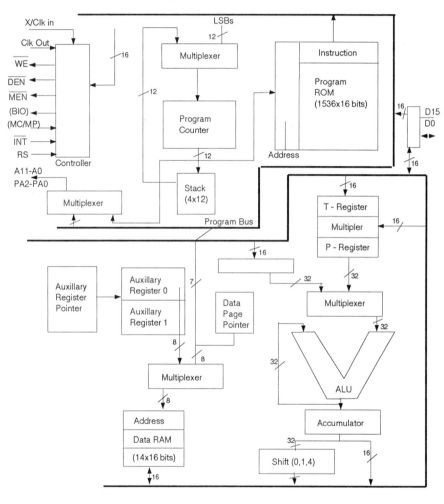

Figure 2 Architecture of Texas Inst. TMS32020/32021 Signal Processor

cution is pipelined. The sequencer contains program counter, two loop counters, external program memory interface, and two instruction registers, which form a pipeline feeding the decoder section, which in turn controls the chip. Using the two instruction registers, instructions are fully decoded and operand addresses are calculated before the AU (Address Unit) starts the actual computation. The addressing section contains two address ALU's so that both operand addresses can be computed in

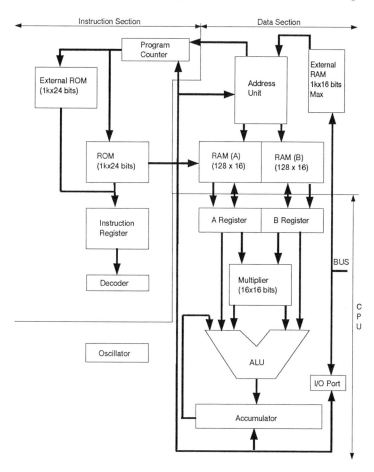

Figure 3 Architecture of Fujitsu 8764 Signal Processor

parallel. It has a 2x128x16 bit RAM. It is divided into two sections to feed the input simultaneously to the ALU. Because of the multiplicity of the functional units, the Fujitsu chip can perform ALU operations in parallel with overhead operations such as address calculations and I/O, thus greatly speeding up the execution time. Instruction word is 24 bits wide. ALU provides a great variety of arithmetic and logical operations including multiply-add and left and right shifts. Another approach to increasing throughput draws on external program and data memories that are fast enough to supply instructions and operands in a single clock cycle. However this requires static RAMs with a minimum capacity of 16384 to 66536 bits and minimum cycle time of 25 to 70ns. By way of comparison, general purpose microprocessors typically use dynamic RAMs with capacity of 262,144 to 1Mbit and external program and data memory cycle times of 250ns to 300ns.

The main recent technological trends are: floating-point processors and RISC processors.

3.1 Floating-Point DSP Processors

They have been the answer to the requirements of various complicated DSP applications which require wide dynamic range of high resolution and precise accuracy. In integer architecture, the result of multiplications and additions must be scaled, because they will quickly go out of range. Operations to perform this type of scaling are complex and take many processor cycles, thus greatly reducing system throughput. In addition, there are many algorithms that are extremely difficult or impossible to program into fixed-point devices. Floating-point chips require fewer support chip to integrate into the system. Some of the trade-offs that must be considered in designing A DSP chips incorporate floating-point math capabilities, they use 16, 18, 22, 24 bit mantissas, and 4, 6, or 8 bit exponents to achieve the best precision/chip-cost trade-offs. However, 32–bit floating point schemes (24 by 8) are the most popular. As a case study, consider the Motorola DSP56000 family. Its performance peaks at 40 MFLOPS. Its also executes 13.33MIPS and can crank out a 1024–point, complex fast Fourier transform in less than 2 ns.

The 96000 chip employs parallelism and concurrency. Its Harvard architecture is made up of 5 data paths and three address buses all 32–bit wide, which link the various on-chip resources. External memory space is addressed through a 32–bit address driven by a 3–bit input multiplier that can select one of the internal address buses. Three 32–bit execution units data ALU, address-generation unit, and program controller operate in parallel with a CPU. An instruction fetch, up to three floating-point operations, two data moves, and two address port updates can be executed in one instruction cycle. Two on-chip DMA controllers assure an adequate supply of data available for processing Fig. 4. Number crunching is handled by a data ALU section which contains a general purpose register file of ten 96–bit registers, floating-point multiplier and adder/subtracter, and a 320 bit barrel shifter. Floating-point ALU results are always 96 bits, and integer results are either 32 or 64bits. The

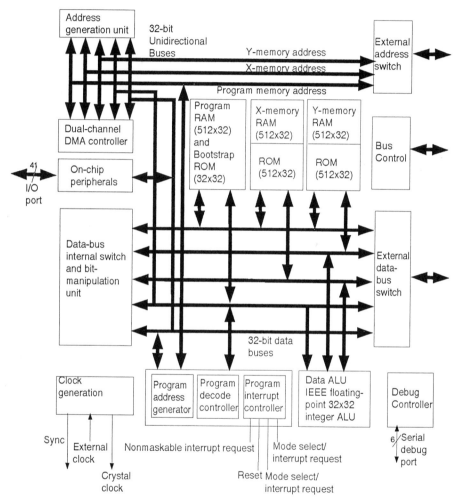

Figure 4 The 9600 Chip Architecture

multiplier supports single-precision or single extended-precision floating-point number format with up to 32–bit mantissa and 11–bit exponents. The adder/subtracter can generate the sum and difference of the same two operands in one cycle, useful for calculating fast Fourier transforms. Shifting operations for normalizing floating-point numbers and scaling integers is the barrel shifter's job. The eight buses, CPU, and 2 DMA's share the chip with six data and program memories (including a bootstrap ROM); three MCU-style peripherals (serial communications interface, synchronous serial interface, and a 32–bit host interface); and a clock generator. The 1.0μm CMOS chip performs a 1025 FFT in only 1.67 ms, with all code and data contained within on-chip memory. An FIR filter tap with data shift takes 60ns, and just 60ns per vector element is needed to complete a vector dot product. Several similar chips are available in the market such as: TI TMS320C30, Fujitsu MB86232, Oki MSM6992, and Intel 80960KB.

3.2 Parallel DSP Chips

The performance of floating point chips are considerably enhanced by including parallelism and pipelining. The 32–bit IQMAC multiplier accumulator chip (from United Technologies Microelectronics Center) is a vector processor. It incorporates three 32–bit floating-point ALU's and two 32–bit floating-point multipliers on one CMOS IC, Fig. 5. Multistage pipelining is also included. The architecture is flexible in the way that it lets the designers choose between FIFO and RAM based memory configurations. With data and control bits pipelined throughout the chip's design, each of the two math sections targets specific functions. The upper math portion — the two multipliers and upper ALU performs complex number multiplications in two cycles. A dual IQMAC system executes the same operation in only one clock cycle by separating the real and complex number calculations.

The lower math section — right and left ALU's plus an accumulator can execute FFT butterflies in two cycles or a complex number accumulation in one cycle. A 1K-point complex FFT takes only 510μs at 20MHz. Using a dual-chip setup, it will take only 341μs. The chip can be easily cascaded either vertically or horizontally for a linear improvement in performance. In a cascaded array, a dedicated chip executes each butterfly pass in an FFT. A 10–chip 1K-point FFT executes in a mere 64μs. The IQMAC chips has six 32–bit ports, including four that are bidirectional. Multiple ports eliminates the data transfer bottlenecks common in one-chip DSP devices, because multiple data words move on and off the chip, simultaneously.

3.3 RISC DSP Chips

DSP processors are generally considered as a special subset of RISC architectures dedicated to performing math-intensive algorithms DSP chip are totally dedicated to providing a multiply/accumulate in a single instruction cycle. As conventional RISC chip continue to mature, there will be less and less differentiation between DSP and RISC processors. Intel 80860 chip is a good example of the merge between both technologies. It comprises two processing unit handles integer operations while the other handles floating-point and graphic operations. The two

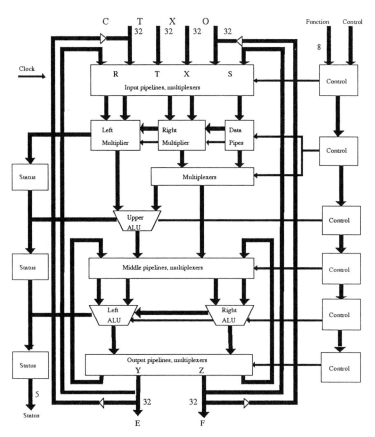

Figure 5 The IQMAC Chip Architecture

can operate independently or work together to provide some high-power DSP-class performance. This high parallel architecture and tight coupling of the integer and floating point unit enable the chip to generate a new result each cycle. Operating at 40MHz, the chip can perform at 40 integer

MIPS and 80 single-precision MFLOPS. Its floating-point units, the FPU adder and FPU multiplier, operate in parallel and exchange data between each other. Unlike conventional RISC processors. 80860 does not have a fixed number of stages. It is a set of pipelined stages.

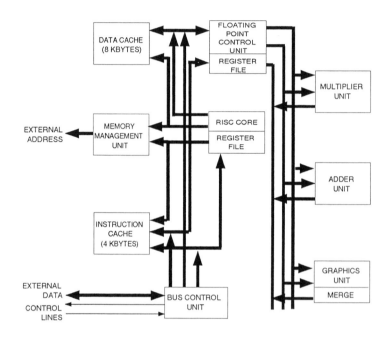

Figure 6 The 80860 RISC Processor

4. DSP CORES

The DSP cores can be considered as a compromise between the flexibility and average performance of certain applications of general purpose processors, and the long design turnaround time, limited design flexibility, and restricted usability of the dedicated processors. The use of programmable processors cores as building blocks allows high levels of flexibility with minimal design efforts. The core processors can be configured to realize different specifications in data rates, memory capacity power consumption, throughput and other processing parameters [4,12]. The design flexibility can be further increased by parameterizing the core processor. Two main approaches can be identified in employing the core, generic core processors and DSP controllers.

4.1 Generic Core Processor

An example of a generic DSP core is shown in Fig. 7 [5]. Two types of ALU are available: a single ripple carry ALU and a faster carry look-ahead (CLA) one. The ALU is parameterized, it can be interfaced to an accumulator, which communicates with other devices over the lens. The shifter is a zero to N-1 bits right barrel shifter. In the register file, there can be four types of registers. The hardware multiplier is not included to minimize the silicon area since digital filters can be designed to use only a limited number of shifts and adds instead of multiplication using full word length. Two RAM blocks are provided, each of 1024 words. They can be configured as delay lines.

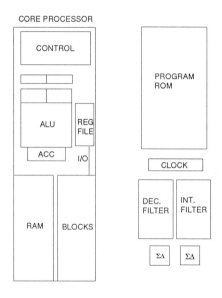

Figure 7 The Generic Core Processor

Baji et al, [4] have presented a 20ns CMOS (1μs) DSP core for video applications. This core processor has a reconfigurable high speed data path supporting several multiply/accumulate functions including 16–tap linear-phase transversal filtering, high speed adaptive filtering, and high speed discrete cosine transform (DCT). Multi-DSP cores can be used, too.

The DSP core consists of an 8–bit x 8–bit modified Booth Parallel Multiplier (MLT), two 12–bit arithmetic logic units (ALUA and ALUB), two sets of 2x8x12–bit accumulator arrays (ACCA and ACCB), and 8–bit

14

x 16–word coefficient memory (CM), and a 4–bit x 64–word coefficients address memory (CAM), Fig. 8. A programmable CMOS phase-locked loop (PLL) circuit is provided as a clock pulse generator high speed operation. It has 40–bit x 8–word microprogram memory. To implement a linear-phase FIR filter, the core will be configured as shown in Fig. 9. One tap of the filter is processed in 10ns. The core will be configured as shown in Fig. 10. To implement a discrete cosine transform (DCT), one point DCT will be processed in 160ns.

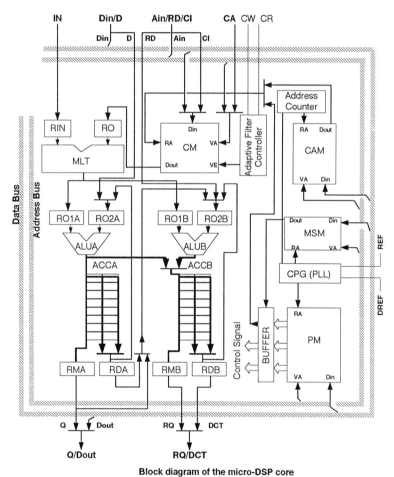

Block diagram of the micro-DSP core

Figure 8 ADSP Core [6]

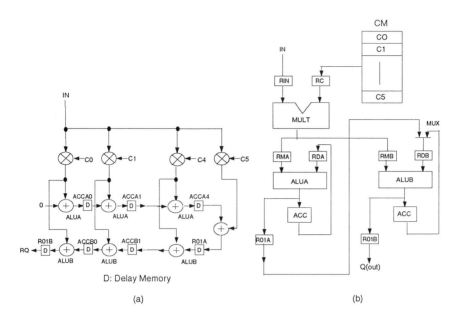

Figure 9 A DSP core configured as a FIR filter (a) Signal Flow Graph (b) Architecture

4.2 DSP controllers

Through the single-chip digital signal processors have achieved an impressive level of integration, they still have to be supported by a host processor supporting the chip, coordinating data transfer, performing other data transfers, and performing other timing and control operations. The main idea of the DSP controllers is to include a controlling module within a DSP core. This architecture represents a revolutionary change in the way engineers will look at control functions. Many of them continue to use analog circuitry because the DSP solution has not been available in a cost-effective way. One promising application area for DSP controllers is in closed loop control where a feedback loop is used to improve control accuracy by compensating for system characteristics. Traditionally, these compensation loops have been built using analog circuitry. Using digital techniques, such control loops are generally implemented using second order FIR and IIR filters. These digital approaches bring higher reliability, noise immunity and flexibility to such applications. Conventional microcontrollers lack the hardware arithmetic capability and cannot compute such filter algorithms in anywhere near real time.

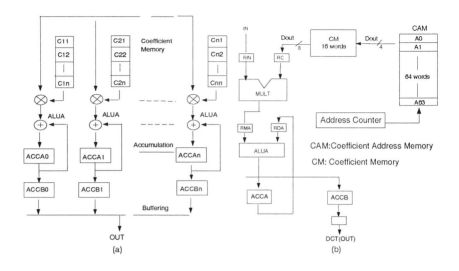

Figure 10 DSP Core connected as a DCT (a) Signal Flow Graph (b) Architecture

As a case study, consider the 320C14 chip (developed by Microchip and TI). It is composed of 16–bit 320C10 DSP core, a high speed serial port for communications or serial data transfer, four times, 16 interrupts signals, 16 latched I/O lines, a timer-based event manger, 6 pulse-width-modulated outputs, 512 bytes of RAM, and 8192 bytes of ROM, Fig. 11. An additional 8192 bytes of off-chip memory can also be addressed. The fast CPU cycle of the DSP core combined with its hardware multiplier make it possible for the 320C14 to execute single cycle instructions such as 16–bit multiply and accumulate operations in 160ns. As a result, the chip can achieve a peak performance of about 6 MIPS when operating at 25.6MHz.

The on-chip resources are employed like active I/O subsystem performing operations such as bit manipulation and event monitoring in parallel with the CPU. These capabilities will allow the chip to perform tasks such as Kalman filtering, adaptive control, system data control, and many other closed loop applications.

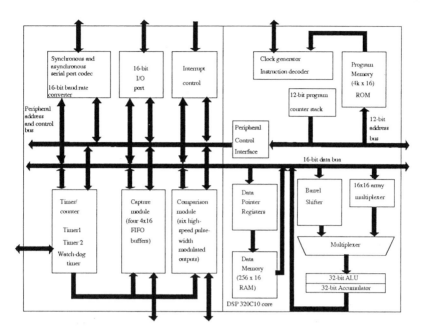

Figure 11 320C14 DSP Controller with 320C10 Core

5. APPLICATION SPECIFIC DSP CHIPS

ASIC DSP chips are developed for specific applications where the required performance cannot be met by general purpose chips. Most of the chips are designed using custom implementation or semi-custom design based on standard cells or gate arrays. The main factor determining the efficient approach is the required volume. With the recent developments in technology, the trend is to use a DSP core as a building block cell supported by other standard cells. AT&T have used DSP 16 as a core processor to be compatible with their CMOS standard cell library. This will let users build standard cells around the core processors and develop a custom part within a reasonable turnaround time. TI has also followed the same approach, 320C25 is used as a core processor surrounded with a gate array configuration.

Further development may see complete DSP macrocells included in many ASIC libraries, this will lead to adopt ASIC as cost effective approach for high volume applications. Algorithm-specific DSP's are more than general purpose chips. Such chips are the implementation of the main DSP and image processing kernels such as finite impulse response (FIR) or infinite impulse response (IIR) filters. An algorithm-specific chip perform differently when it is used in different applications.

Application-specific DSP chips may have a lot features similar to most of the general purpose DSP chips. The techniques discussed for achieving throughput and speed are also employed in ASIC architecture design. These chips amy also be programmable in a very narrow sense such as changing the coefficients of computation etc, but they are not programmable for a completely different problem, the strategies adopted for achieving high speed are usually ad hoc and cannot be generalized for other algorithms.

The special purpose architectures for algorithms like Fast Fourier Transform, Median Filter, IIR Filter, Convolution, Hadamard Transform, image coding etc. are directly derived from their computational structures. For some of the algorithms, the behavior is expected in terms of already derived structures and hence a mapping at the programming level is sufficient. As an example, let us consider the DSP implementation of FFT chip. The order of computation is as shown in Fig. 12 [6]. The architecture developed for this type of computation is as shown in Fig. 13 [7]. It consists of three pipelined processors which work for the three stages of computation. The architectural structure has features which is a natural outcome of the computational structure. Algorithm specific DSPs have different architectural features than those of general purpose ones in many respects. To highlight a few, the bus structures are not regular and may have many multiple and non-uniform buses. The control unit is not explicitly hardwired or microprogrammed but a combination of the two. Memories have multiple ports and are usually not of standard sizes. The architecture is more ad hoc rather than following a thumb rule. Some of the architectures may have vector functional units for matrix or vector multiplications. Many commercial chips which are algorithm specific are available now in the commercial market. Examples are the interpolation filter chip used in the compact disc players, modems and CCITT standards etc. Fig. 14 shows an algorithm specific architecture for discrete cosine transform (DCT) [8], while Fig. 15 is an ASIC architecture for Kalman filter [9].

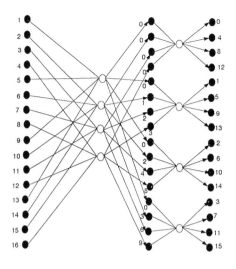

Figure 12 Computational Structure of a 16–point radix-4 FFT

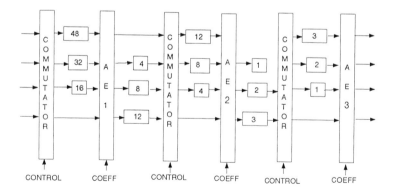

Figure 13 Architecture of the radix-4 16–point FFT processor

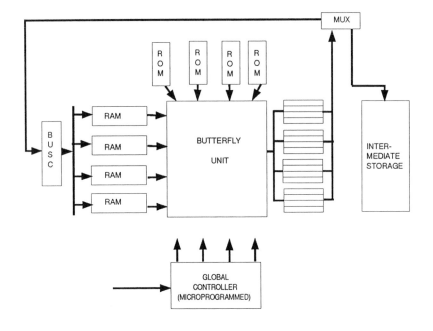

Figure 14 An ASIC Chip for DCT

6. EVOLVING DSP APPLICATIONS

The rapid developments in VLSI technology has led to wide new applications for VLSI DSP architectures. To new applications which are discussed here are; Sensors and High Definition TV (HDTV).

6.1 Sensors

Sensors prototyping have gained great significance recently. Smart sensors can be realized by integrating sensing and signal processing. DSP processors have been employed as the computational kernel in applications such as: computer vision, robotics, and consumer products. With smart sensors, one can acquire 100 to 1000 frames of images per second. With the advances in technology, a 312x287 pixel sensor array, together with all the necessary sensing and addressing circuitry can be realized in 9x9 mm using 2μm, 2–level metal CMOS technology. As an example, consider the application of image scanners [11]. One main function is obtaining color transformation. It can be achieved by using three linear arrays, each with a different color filter (R, G, or B), integrated on the surface of the sensor. These sensors are spaced apart by an integral number of lines on the same substrate. The page scan direction is perpendicular to the linear arrays.

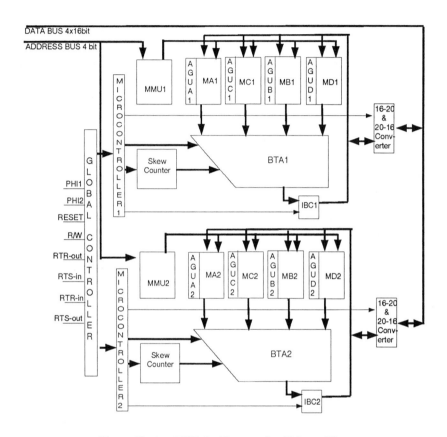

Figure 15 An ASIC Architecture for Kalman Filter

The design of the DSP-L module has been reported in [11]. The chip can perform the following functions: Digital correlated, double sampling, black level corrections and gain correction for each photosite, sensor defect concealment, color matrixing, and look-up table operations for space transformations. Additionally, three interface circuits are implemented to store and retrieve black-level and gain correction values, rephase red, green, and blue values to provide line coincidence, and function as a simple computer lens surface, for loading coefficients and writing the processed image data. A prototype has been built using 2μm CMOS technology. It contains 132,000 transistors in 13.6x11.3 mm area.

6.2 High Definition TV

High Definition TV is being developed as the next-generation television standard. It has a 16.9 width-to-height ratio and more than 1000 scan lines in a frame, which produce approximately five times as much information as current TV systems. Such improvements represent new technological challenges. Sophisticated real-time signal processing and compression techniques are required. Large amounts of memory and specialized logic is needed. HDTV will help in building new technologies such as multimedia workstations. VLSI chips required for HDTV are distinguished from traditional video signals. HDTV systems must process about 60 million pixels/sec in real time. Most existing HDTV productions and transmission schemes are analog or mixed digital/analog. It is expected that digital techniques will lead to good picture enhancement in a cost-effective way. Data compression algorithms, techniques, and architectures have gained considerable significance for HDTV broadcasting and receiving. The net bit-rate generated by uncompressed HDTV is approximately 1 Gbit/sec. It is believed that digital "contribution quality" HDTV i.e., program material sent from one studio to another, should be in the 200–300 Mbits/sec range to permit further editing. On the other hand, program material sent to the home should be in the range of 45–100 Mbits/sec.

Function	Multiplication per sample
Single Multiplication	1
L2 Norm Calculation	1
3 x 3 Spatial Filter	9
Ten-Stage Binary Tree Search Pattern Matching (Vector Quantizer with 1024 vectors)	10
2D Matrix Transform (DCT) (8 x 8 kernels)	16
Three-Stage Octal Tree Search Pattern Matching (Motion Vector Detection)	25

Video compression represents one of the main video processing tools which is also computation intensive. As an example, Table 1 [10] lists the number of the required multiplications per sample for several Video Processing functions. Discrete Cosine Transform (DCT) has been employed for this tasks. In DCT, the image is first decomposed into square blocks of typical size 8x8, a DCT transform is applied on each block to produce another 8x8 block, whose coefficients are then quantized and coded. Compression is achieved because most of these coefficients are zeroes. An approximate version of the original block can be generated by applying inverse DCT. This approach requires heavy computation, for example, a DCT based motion compensation interframe encoder composed of a motion detector, DCT, inverse DCT, and an inner loop filter must perform about one billion multiplications per second.

References

[1] Amnon Aliphas and Joel A. Feldman, "The Versatility of Digital Signal Processing Chips," IEEE Spectrum, June 1987, pp. 40-45.

[2] B. Sikstrom, L. Wanhammar, M. Afghahi, and J. Pencz, "A High Speed 2–D Discrete Cosine Transform Chip," Integration, the VLSI Journal, May 1987, pp. 159-169.

[3] S. Y. Kung, "On Supercomputing with Systolic/Wavefront Array Processors," Proc. IEEE, vol. 72, No. 7, July 1984.

[4] T. Baji, et al., "A 20ns CMOS Micro-DSP Core for Video-Signal Processing," IEEE J. Solid State Circuits, Vol. 23, No. 5, Oct. 1988, pp. 1203–1211.

[5] Jari Numi, et al., "Parametrized Macro Module Generation for Digital Signal Processing," Proc. VLSI Signal Processing IV, 1990, pp. 13-22.

[6] Z. Wang, "Fast Algorithms for the Discrete W Transform and the Discrete Fourier Transform," IEEE Trans. ASSP, Aug. 1984, pp. 803-816.

[7] M. Vetterli and H. Nussbaumer, "Simple FFT and DCT Algorithms with Reduced Number of Operations," Signal Processing, Vol. 6, No. 4, Aug. 1984, pp. 267-278.

[8] Srinivas Subramaniam, "VLSI Architecture for a Discrete Cosine Transform," Master's Thesis, The Center for Advanced Computer Studies, Univ. of Southwestern Louisiana, Fall 1991.

[9] Padma Akkiraju, "An Algorithm Specific VLSI Architecture for Kalman Filter," Master's Thesis, The Center for Advanced Computer Studies, Univ. of Southwestern Louisiana, Spring 1990.

[10] Kunihiko Niwa, et al., "Digital Signal Processing for Video," IEEE Ckts. and Devices Magazine, Vol. 6, No. 1, Jan. 1990, pp. 27–33.

[11] W. A. Cook, et al, "A Digital Signal Processor for Linear Sensors,"Proc. ICCC, 1990.

[12] Uramoto et al, "A 100 MHz 2–D Discrete Cosine Transform Core Processor", IEEE Journal of Solid State Circuits, Vol. 27, No. 4, April 1992, pp. 492-499.

[13] M. A. Bayoumi, editor, Parallel Algorithms and Architectures for DSP Applications, Kluwer Academic Publishers, 1991.

2

RECENT DEVELOPMENTS IN THE DESIGN OF IMAGE AND VIDEO PROCESSING ICs

Konstantinos Konstantinides and Vasudev Bhaskaran
Hewlett-Packard Laboratories

1. Introduction

Hardware support for image processing is very important in emerging applications such as desktop publishing, medical imaging, and multimedia. In the past, computational needs for intensive image processing tasks, such as image analysis of satellite data and pattern recognition, were satisfied with custom, complex, and expensive image processing architectures [1], [2]. However, the latest scientific workstations have enough compute power for low-cost desktop image processing. Furthermore, traditional image processing operations, such as texture mapping and warping, are now combined with conventional graphics techniques. Hence, there is an increased interest for accelerated image and video processing on low cost computational platforms, such as personal computers and scientific workstations.

Many developers provide already some type of image processing support, based mostly on general purpose microprocessors or digital signal processors (DSPs), such as the i860 from INTEL, or the TMS320 family from Texas Instruments. However, new application areas, such as high-definition TV (HDTV) and video teleconferencing demand processing power that existing general purpose DSPs cannot provide.

Until recently, commercially-available image processing ICs performed only low-level imaging operations, such as convolution, and had very limited, if any, programming capability. This was due to the well defined operation and data representation of the low-level imaging functions; furthermore, there was a general feeling that the image processing market was quite small. Emerging standards, such as JPEG (Joint Photographic Experts Group) and MPEG (Moving Picture Experts Group) for image and video compression, and the opening of new market areas, such as multimedia computing, make it easier now for manufacturers to invest in the design and development of a new generation of image and video processing ICs. For example, image compression ICs are now becoming widely available. Thus, a new generation of general purpose image processors will soon emerge to extend the capabilities of general purpose DSPs.

In this paper we present a brief overview of the specific requirements in image processing and some recent developments in the design of image and video processing ICs. Both application-specific (ASIC) and programmable image processors are discussed. For the application-specific processors, special emphasis is given to the processing requirements of recent standards in image and video compression. We close with a discussion on a "generic" general purpose image processing architecture. As general purpose DSPs share many common features (a Harvard architecture, multiple data memories, etc.), we expect that future general purpose image processors will share many of the features embodied in the generic design.

2. Image Processing Requirements

Image processing is a broad field that spans a wide range of applications such as document processing, machine vision, geophysical imaging, multimedia, graphics arts, and medical imaging. A careful examination of the imaging operations needed for these applications suggests that one can classify the image processing operations into low, intermediate, and high levels of complexity. In low-level processing (i.e. filtering, scaling, and thresholding), all operations are performed in the pixel domain and both input and output are in the form of a pixel array. In intermediate-level processing, such as edge detection, the transformed input data can no longer be expressed as just an image-sized pixel array. Finally, high-level processing, such as feature extraction and pattern recognition, attempts to interpret this data in order to describe the image content.

Because of this large range of operations, it seems that every conceivable type of computer architecture has been applied at one time or another for image processing 2 and that so far, there is no single architecture that can efficiently address all possible problems. For example, Single Instruction Multiple Data (SIMD) architectures are well suited for low-level image processing algorithms; however, due to their limited local control they cannot address complex, high-level, algorithms [3]. Multiple Instruction Multiple Data (MIMD) architectures are better suited for high-level algorithms, but they require extensive support for efficient inter-processor communication, data management, and programming. Regardless of the specifics, every image processing architecture needs to address the following requirements: processing power, efficient data addressing, and data management and I/O.

2.1 Processing Power

To better understand the computational requirements of image processing algorithms, consider first the 1-D space. In 1-D digital signal processing, a large class of algorithms can be described by

$$Y(i) = \sum_{k=1}^{p} C_k X(i - k), \tag{1}$$

where $\{X(i)\}$ and $\{Y(i)\}$ denote the input and output data, and $\{C_1, C_2, \ldots, C_p\}$ are algorithm-dependent coefficients. Using a general purpose DSP that has a multiply-accumulate unit and can access simultaneously the input data and the coefficients, an output sample can be generated every p cycles.

In image processing, a large class of algorithms can be described by

$$Y(i, j) = \sum_{m=1}^{p} \sum_{n=1}^{q} C_{mn} X(i - m, j - n), \qquad (2)$$

where X and Y are the input and output images, and C is a $p \times q$ matrix of algorithm-dependent coefficients. From (2), a single DSP requires now at least pq cycles to generate an output pixel. In a multi-DSP system, at least q DSPs have to be used to generate an output sample every p cycles, provided that all processors have direct access to the data and there exists efficient inter-processor communication and a data distribution mechanism. Thus, the first main requirement in image processing is *processing power.*

2.2 2-D Addressing

Another major requirement in image processing is *efficient data addressing.* In conventional DSPs, a simple p-modulo counter is adequate for the data addressing in (1) [4]. However, in image processing, the addressing schemes may be far more complicated. For example, Fig. 1 shows an $N \times M$ image, and a $p \times p$ kernel of coefficients positioned with its left upper corner at address A of the image. From (2), to compute an output sample, one possible scheme to sequentially access the image data is to generate data addresses in the following sequence.

$$
\begin{array}{llll}
A, & A + 1, & \cdots & A + (p - 1), \\
A + N, & A + N + 1, & \cdots & A + N + (p - 1), \\
A + 2N, & \cdots & \cdots & A + 2N + (p - 1), \\
\cdots & \cdots & \cdots & \cdots \\
A + (p - 1)N, & A + (p - 1)N + 1, & \cdots & A + (p - 1)(N + 1).
\end{array}
$$

Fig. 2 shows a 2-D address arithmetic unit (AAU) as implemented on the Video DSP by Matsushita [5]. This AAU has two parts, the controller, and the address generator. The AAU operation is controlled by the values in five registers: SA, NX, NY, DX, and DY, as shown in Fig. 2 . In our case, SA=A, NX=p, NY=p, DX=1, and DY=$N - (p - 1)$. To generate addresses from A to $A + (p - 1)$, the adder is incremented by DX=1. After each row of the kernel is processed, the "Row End" signal enables DY=$N - (p - 1)$ to access the adder, and an address is generated for the beginning of the next row. Operations continue until the "End of Block" signal is enabled.

The HSP45240 IC from Harris Semiconductor is a programmable 24-bit address sequence generator for image processing applications. It can be configured to generate addresses for filtering, FFT, rotation, warping, zooming, etc. The Harris IC is ideal for block oriented address generation. Five configuration registers allow a user to define such parameters as: the beginning address of a sequence, the block size,

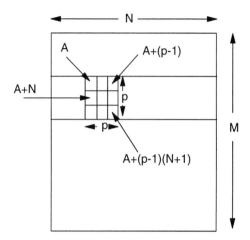

Fig. 1 : 2-D addressing in image processing.

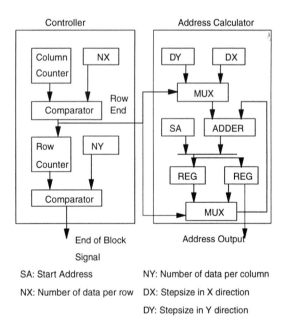

Fig. 2 : Block diagram of a 2-D address generator.

address increment, number of blocks, etc. A crosspoint switch can be used to reorder the address bits. This allows for *bit-reverse* addressing schemes used in FFT computations.

Recent image processor ICs have very long instruction words (VLIW) for the control of multiple AAUs, ALUs and function units. A programmer has to simultaneously control not only the two to three address generators, but also all the other function units. Kitagaki et al. [6]. presented an AAU design that allows the user to control the address generator controller using high-level code words. Given the high-level code word, an address microsequencer then generates the full sequence of addresses required in a particular application. The design allows for 17 different high-level and commonly used image processing addressing modes, such as: block raster scan, neighborhood search, 1-D and 2-D indirect access, and has special modes for FFT and affine transformation addressing.

2.3 Data Storage and I/O

A 512 x 512, 8-bit grayscale image requires 262 Kbytes of storage and a 24-bit, 1024 x 1024 color image requires approximately 3 Mbytes of storage. Considering that many applications require data processing in real-time (12-30 frames per second or more), it is easy to see why *efficient data storage and I/O* are extremely important in image processing architectures.

Systolic architectures address this problem by distributing the data across an array of processors and allow only local data transfers. Many DSPs achieve efficient I/O by integrating a local Direct Memory Access (DMA) engine with the rest of the architecture 4; furthermore, I/O operations and computations are usually performed in parallel. As we will see next, similar techniques can also be applied into the design of image processor ICs.

Image processing architectures can be divided into the following broad categories: dedicated image processor systems, image processing acceleration systems, and image processor ICs. The dedicated image processor systems usually include a host controller and an imaging subsystem that includes memory, customized processing ICs, and custom or commercially-available math processing units, connected in a SIMD, MIMD, or mixed-type architecture. Most of these systems are being developed either at universities for research in image processing architectures, or at corporate laboratories for in-house use or for specific customers and application areas 3, [7], [8].

Image processing acceleration systems are usually being developed as board level subsystems for commercially-available personal computers or technical workstations. They use standard bus (VME, EISA, NuBus, etc.) interface to communicate with the main host, and usually include memory, one to four commercially-available DSPs (such as the TMS320C40 from Texas Instruments) or high performance microprocessors (such as the INTEL i860), a video port, and a frame buffer. Most of the board-level subsystems also include floating-point arithmetic unit(s) and have a Bus-based architecture [9].

In this paper, the emphasis will be on image processor ICs. These are either application-specific circuits (for example, image compression ICs) with limited programming capability, or programmable circuits designed for a specific range of applications (HDTV, video conferencing, etc.). No single article can cover all existing designs. Hence, we present only a selection of chips that cover the current trends in hardware for image and video processing.

3. Application Specific Image Processor ICs

Most image-processing functions can be implemented on a general purpose digital signal processor and several vendors offer such solutions today. Such solutions are acceptable to test the functionality of a particular image processing algorithm. However, they are generally not amenable to real-world applications due to the processing requirements of those applications and the resulting high cost associated with the use of a multi-DSP system.

For example, in computer-vision systems, one might want to perform a semi-automated PC board inspection. This task requires imaging a PC board at video rates (256 x 256 x 8 bits/pixel acquired at 30 frames/sec), and performing edge-detection on the digital image followed by binarization of the edge-detected image. Then, a template matching function might be performed on the binary image to determine faulty traces on the board. The above-mentioned tasks require at least 120 million operations per second of processing capability. Most general purpose DSPs are not capable of delivering such a high MOP count at sustained rates. To make applications like these feasible, IC vendors offer single-chip solutions for many of these image processing tasks.

Image compression is another area where a generic DSP solution is not acceptable. With the emergence of multimedia computing and low-cost imaging peripherals such as color scanners and color printers, there is a large body of emerging applications that will incorporate high-quality images and/or digital video. To facilitate the delivery and storage of high-quality images and digital video, considerable effort is being expended in developing compression standards for such data-types. Presently, the JPEG compression standard for still-images and the MPEG compression standard for digital-video (primarily for CD-ROM based applications) have motivated several IC vendors to develop single-chip solutions conforming to these standards. Desktop video conferencing applications have also motivated IC vendors towards offering custom chip solutions conforming to the Px64 (H.261) video compression standard. It is envisioned that workstation based computing systems in the next few years will contain functionality offered by such chips and this will enable the workstations to handle still-images and video in a real-time manner. In this section we will describe application-specific imaging ICs suitable for low-level image processing and image compression.

3.1 Image Processing ICs for Computer-Vision Systems

In computer vision and most traditional image processing systems, a significant amount of processing is performed with low-level image processing functions, such as image enhancement and edge-detection. For such applications there exist several high-performance ICs.

ICs for Image Filtering

LSI Logic[10] offers the L64240 multi-bit finite impulse response filter (MFIR). This chip is a transversal filter consisting of two 32-tap sections and each 32 tap section is comprised of four separate 8th order FIR filter sections. The output can be generated at a 20 MHz sample rate with 24 bits precision. The data and coefficients supplied to each filter section can be 8 bits wide. Each filter cell within a filter section consists of a multiplier and an adder which adds the multiplier output and the adder output of the preceding filter cell. The multiplier accepts 8 bit inputs (data and coefficients) and the adder accepts 19 bit inputs. The adder output serves as the adder input to the next filter cell within this FIR filter.

A format adjust block at the end of each 32-tap section provides the capability to scale (with saturation), threshold, clip negative values, invert or take absolute value of the output (these functions are useful in image enhancement and in image display). Since the convolution kernel can be loaded synchronously with the device clock, it is possible to perform adaptive filtering and cross-correlation with this device. By changing the kernel (coefficient data), this filter can be used for edge-sharpening (useful as an image enhancement function), noise-reduction (preprocessing step in many imaging applications wherein the image is acquired) and image resampling (for printing).

The L64240 is reconfigurable and can perform 1-D, 2-D and 3-D filtering. By cascading several MFIRs, input data and convolution kernel precision can be extended to 24 bits. This is accomplished by splitting the data sample into its high and low-order bits, performing convolutions separately on these two streams, and combining the partial results. Furthermore, complex filtering can also be accomplished by configuring the MFIR as four 16-tap filter sections and by feeding the real and imaginary parts of the data into these filter sections.

ICs for Image Enhancement

In many imaging applications, the image data is acquired from a noisy source, e.g. scanned image from a photo-copy, and some noise reduction function needs to be applied to this image. One such technique is rank-value filtering, e.g. median-filtering. An IC that performs this task at a 20 MHz rate is the L64220 from LSI Logic 10. Like the MFIR, this is a 64-tap reconfigurable filter. However, the output is from a sorted list of the input and not a weighted sum of the input values, like the MFIR's filtering function. The filter can be configured to perform rank-value operations on 8x8, 4x16, 2x32 or 1x64 masks. The rank-value functions include min, max and the median function. The processor is pipelined and each stage of the pipeline operates on one bit of the input data. The total number of ones in the input data at a certain bit position (in one stage) is counted. If this number exceeds (or equals) the rank specified by the user, output bit for that position is one. A mask (which indicates which of data words might still affect the output) is passed from stage and is modified. The basic operations performed include ANDing (for masking), summing and comparing.

In many image processing applications the acquired image needs to be enhanced for incorporation within an electronic document. Image enhancement in this context implies equalizing the histogram[11] of the image. This function requires computing the histogram of the input image and then applying a function to the input which equalizes the histogram 11. The L64250 IC from LSI Logic performs the histogram operation at a 20 MHz rate on data sets as large as 4K x 4K pixels. Similarly, the HSP48410 "Histogrammer" IC from Harris Semiconductors can perform signal and image analysis at speeds up to 40 MHz.

ICs for Computer-Vision Tasks

The L64250 chip can also perform a Hough Transform. The Hough transform technique transforms lines in cartesian coordinates to points in polar coordinate space11. This property can be used to detect lines in the input image. Line detection is important in many computer-vision tasks such as detecting connectivity between two points on a PC board.

Another IC useful in computer-vision tasks and optical character recognition (OCR) is the object contour tracer (L64290 from LSI Logic). An internal image RAM stores the 'binary' image of up to 128 x 128 pixels. It also stores a 'MARK' bit for each pixel indicating contour points that have been already been traced. This chip locates contours of objects in this image and for each contour, it returns the (x, y) coordinates, discrete curvature, bounding-box, area and the perimeter. In an OCR application, this data can be part of a feature set that can be used to perform template matching in feature space to recognize this character (whose 128x128 image is input to the contour tracker). In conventional computer-vision applications, the contour tracker can be used as a component in a system to identify objects. This could be the front-end signal processing component of a robot. For an NxM image, the contour tracker takes 4.5 NM cycles to generate the contours.

This IC can also be used for the compression of black-and-white images wherein the contour information will represent the compressed data. For most textual documents, this could yield very high compression ratios compared with the traditional Group III and Group IV fax compression methods.

In image recognition tasks, template matching might be employed to identify an object within a scene. In many applications the template matching is performed in image space. To facilitate this process, the L64230 (from LSI Logic) performs binary template matching with a reconfigurable binary- valued filter with 1024 taps. The filter can operate at 20 MHz in various configurations including 1-D and 2-D filtering, template matching, and morphological processing.

ICs for Image Analysis

In this category, we find several ICs that accept spatial domain image data and generate non-image data. Most of the ICs in this class perform some form of frequency domain processing and the FFT is the most popular function. Plessey Semiconductor and LSI Logic (L64280)10 (among others) have high-speed FFT processors that are capable of computing FFT butterflies at a 20 MHz rate.

3.2 ICs for Image and Video Compression

Workstations today offer high performance and support large storage capacities, thus being viable platforms for multimedia applications that require images and video. However, the bandwidth and computation power of the workstations is not adequate for performing the compression functions needed for the video and image data. A single 24-bit image at 1K x 768 resolution requires 2.3 Mbytes of storage, and an animation sequence with 15 minutes of animation would require 60 Gbytes. To facilitate such applications and multimedia computing in general, recently, there has been a great deal of activity in the image and video compression arena. This activity has progressed along two fronts, namely the standardization of the compression techniques and the development of ICs that are capable of performing the compression functions on a single chip.

The image compression standards currently being developed are in four classes as shown in Table 1. The JPEG standard [12] is intended for still-images. However, it is also being used in edit-level video applications, where there is a need to do frame-by-frame processing. The MPEG (1 and 2) standards apply to full-motion video; MPEG-1 was originally intended for CD-ROM like applications wherein, encoding would be done infrequently compared with decoding. MPEG-1 [13] offers VHS quality decompressed video whereas the intent of MPEG-2 is to offer improved video quality suitable for broadcast applications. Recently, video broadcast companies have proposed video transmission schemes that would use an MPEG decoder and a modem in each home. By optimizing the decoder and modem configuration, the broadcasters are attempting to incorporate up to four video channels in the present 6 MHz single channel bandwidth. The increased channel capacity has significant business implications.

The Px64 (H.261) standard [14] is intended for video conferencing applications. This technology is quite mature but the devices available to accomplish this have been until now multi-board level products. As CPU capabilities increase, we expect some of these functions to migrate to the CPU. However, a simple calculation of the MIPS (million instructions per second) requirements for these compression schemes indicates that such an eventuality is at least five years away. For example, in Table 2, we show the MIPS requirements for a Px64 compression scheme. Note that the 1000 MIPS requirement for encoding is not going to be achievable in a cost-effective manner on a desktop in the near future. Recently, several software-only decompression procedures have been developed for MPEG-1 and Px64 real-time decoding on workstations. These decompression procedures rely on efficient partitioning of the decoding process so as to offload some of the computations to the display processor. For the remaining compute-intensive tasks, several fast algorithms

Table 1
Image And Video Compression Standards

FEATURES	JPEG	MPEG-1	MPEG-2	Px64
Full-color still images	Yes			
Full-motion video	Yes	Yes	Yes	Yes
Real-time video capture and playback	Yes			Yes
Broadcast-quality full-motion video	Yes		Yes	
Image size (pixels)	64K x 64K (max)	360 x 240	640 x 480	360 x 288
Compression ratios	10:1 to 80:1	200:1 (max)	100:1 (max)	100:1 to 2000:1
Typical data rates (compressed data)		1.5 Mbps	5-20 Mbps	64 Kbps to 2 Mbps

have been devised; for example, recently, there have been several developments in fast inverse DCT (Discrete Cosine Transform) methods that are well suited for these compression standards. For applications requiring real-time compression and decompression, based on the MIPS requirements discussed in Table 2, it seems that application-specific ICs may be the only viable near-term solution. In the JPEG case, application-specific ICs are useful when dealing with motion sequences. For still-images, it is possible to optimize the JPEG algorithms to accomplish interactive processing on a desktop workstation.

In Table 3, we list representative image and video compression chips that are now commercially available. Similarly, in Table 4, we list representative image and video compression chips that are in a research and/or development stage. In addition to the vendors shown in these tables, other major IC companies, such as Texas Instruments and Motorola are planning to introduce both programmable and application specific ICs (ASICs) for video compression. Recently, these IC vendors have also announced variations on the MPEG chips that are appropriate for a low-cost implementation of decoders for set-top converters. These variations are primarily intended to : a) reduce memory requirements and decoding latencies by eliminating B frames in the MPEG bitstream, b) support the interlaced scanning mode for video, and c) allow fast switching between the video channels.

All of the compression chips have a common set of core functions, namely a spatial-to-frequency domain transformation, quantization of frequency-domain signals, and entropy-coding of the quantized data. For compression of motion sequences, additional processing in the temporal domain is also employed. The main processing flow for JPEG, MPEG, and Px64 is depicted in Fig. 3 . We will briefly describe the main processing functions in these compression schemes.

Table 2

MIPS Requirements for Px64 Compression And Decompression
(Image Resolution: 360x288, YCrCb Encoding, 30 frames/sec.)

COMPRESSION	MIPS
RGB To YCrCb	27
Motion Estimation i.e., 25 searches in a 16x16 region.	608
Coding Mode Motion-vector only mode, Interframe coding or Intraframe coding	40
Loop Filtering	55
Pixel Prediction	18
2-D DCT	60
Quantization, Zig-zag scanning	44
Entropy Coding	17
Reconstruct Previous Frame (a) Inverse Quantization	9
(b) Inverse DCT	60
(c) Prediction+Difference	31
TOTAL	**969**
DECOMPRESSION	
Entropy Coding - decoder	17
Inverse Quantization	9
Inverse DCT	60
Loop Filter	55
Prediction	30
YCrCb to RGB	27
TOTAL	**198**

Spatial-to-frequency domain transformation: This is accomplished via an 8x8 2-D DCT (Discrete Cosine Transform) performed on the spatial domain data. Often the RGB spatial domain data is first converted into a color space suitable for image compression. The YCrCb color space is used for this purpose. The 2-D DCT can be performed as a straightforward matrix multiply operation or as a matrix-vector multiply operation (on 16 vectors), or via a fast DCT algorithm. Straightforward matrix-vector multiplication methods are well suited for a hardware implementation whereas a programmable image computing engine (such as a DSP) might use a fast DCT implementation.

Quantization: The frequency domain samples are quantized so that some of the

Table 3
Commercially Available
Image And Video Compression ICs

Vendor	Part No.	Standard	Comments
C-Cube Microsystems	CL550 CL450 CL950	JPEG MPEG-1 MPEG-1	10 and 27 MHz Decode only. Can be used at MPEG-2 resolutions.
	CL4000	JPEG/MPEG/ Px64	Encode,Decode
AT&T	AVP-1300E	JPEG/MPEG-1/ Px64	Encoder only. Data rates up to 4 Mbps.
	AVP-1400D	JPEG/MPEG-1/ Px64	Decoder only. Spatial resolution up to 1K x 1K.
Integrated Information Technology	VP	JPEG/MPEG-1/ Px64	MPEG-1 encoding at non-real time. Real-time decoding at 352x288 spatial resolution (CIF).
LSI Logic	LS647xx	JPEG/MPEG-1/ Px64	Requires up to 7 ICs for multistandard compression.
	L64702 L64112	JPEG MPEG	240x352 pixels. Video decoder.
SGS-Thompson	STV3208	JPEG/MPEG	Performs 8x8 DCT for JPEG and MPEG core.
	STI3220	MPEG/Px64	Performs motion-estimation for MPEG and Px64 encoding.

frequency components can be eliminated. This results in a loss of detail which worsens as the compression ratio is increased. The quantization function is a simple pointwise divide (and rounding) operation.

Entropy-coding: The quantized data has few values and there is usually a run of zero-valued samples between non-zero valued samples. A Huffman coding method employed on the runs is usually used (e.g. Baseline JPEG standard).

Table 4
Image And Video Compression ICs
in R & D stage

Vendor	Standard	Comments
Matsushita	H.261/JPEG/MPEG	H.261 15 frames/sec. 64 Kbps encoder output rate. 352x288 spatial resolution.
NEC	JPEG/MPEG-1/ Px64	352 x 240 encoding at 30 frames/sec. JPEG solution requires only two chips.
Pioneer	JPEG/MPEG-1/ MPEG-2/H.261	25 Mbits/sec (max), Three-chip set decodes only.
Array Microsystems	MPEG-1/Px64 JPEG	2-chip prog. set. MPEG encode for SIF, JPEG encode/decode. (720x480), MPEG for NTSC.
Integrated Information Technology	JPEG,MPEG,Px64	MPEG-2 encoding at CCIR 601 resolution is in non real-time.
Media Vision	Motive video encode/decode. 1:16 max	640x480 pixels not MPEG compliant.

Motion-compensation: In CD-ROM applications, since a CD-ROM drive's output data rate is usually 1 - 1.5 Mbits/sec and digitized NTSC rates are at 80 - 130 Mbits/sec, compression ratios around 100:1 are sought for CD-ROM based video. The MPEG-1 standard was motivated by the expectation that many video applications will be CD-ROM based, e.g. computer-based training. To achieve such high compression ratios, one needs to exploit both spatial and temporal redundancies. This is accomplished through *motion-compensation*. This technique is used in MPEG and Px64 encoding, wherein, frame-to-frame changes are determined and only the changes are transmitted. In most implementations, some form of block-matching is performed between blocks (frames are partitioned into non-overlapping blocks). This is highly parallelizable and this feature is exploited in some of the MPEG and Px64 encoders such as the CL4000 from C-Cube Microsystems. Next, we will briefly describe some of the compression ICs offered by various vendors.

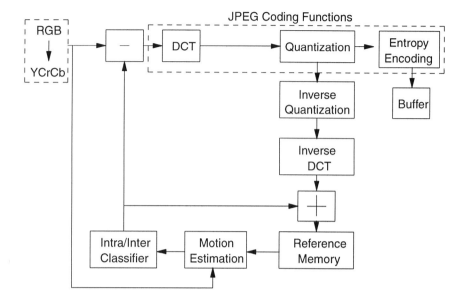

Fig. 3 : Main processing flow in JPEG, MPEG, and Px64 coding schemes.

JPEG - CL550

The CL550 is a single-chip JPEG compression and decompression engine offered by C-Cube Microsystems. This chip is depicted in Fig. 4. Compression ratios can range from 8:1 to 100:1 and are controlled by the quantization tables within the chip. The chip has on-chip video and host-bus interfaces and thus minimizes the glue logic needed to interface the chip to the computer's bus or to the display sub-system's bus. The chip can handle up to four color channels and thus can be used either in the printing environment (images are handled as CMYK) or in the image display and scanning environment (images are handled as RGB). Memory interface signals are also provided by the chip. Since the chip can process data at 27 MHz, it can be used in full-motion video applications wherein frame-to-frame access is needed (MPEG uses interframe coding methods which require several frames to be decoded in order to access a specific frame). Similar chips are presently being offered (or proposed) by SGS-Thompson, and LSI Logic.

MPEG - CL450

The CL450 is a single-chip MPEG-1 decoder from C-Cube Microsystems. Compact disc systems which play back audio and video use such a chip for video playback. Like the CL550, glue logic required for interfacing to busses and other chips is minimized by providing bus logic and memory control signals within the chip.

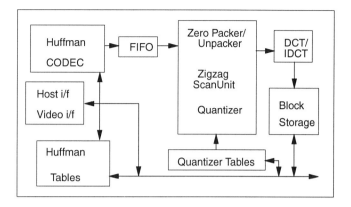

Fig. 4 : Block diagram of the CL550.

3.3 Combined Chip Sets

Multimedia computing systems will have to deal with JPEG, MPEG and Px64 compressed data. Computing systems with multimedia authoring capabilities will also generate data conforming to these standards. However, for cost-effective solutions, it is highly unlikely that separate chips will be used to implement each of these standards (which, as shown in Fig. 3, share many common operations). Thus, several IC vendors have developed (or proposed) JPEG/MPEG/Px64 chip designs capable of handling all three of the major multimedia compression, decompression and transmission standards.

AT&T Multi-standard Chipset

AT&T has announced three chips which could handle all of the multimedia formats. The AVP-1400-C system controller addresses the issue of transmission of multimedia data (this is a key differentiator in AT&T's multimedia chipset offering). The controller provides multiplexing/demultiplexing of MPEG or Px64 streams and performs error correction. Furthermore, via its concentration highway interface (CHI) provides an interface to channels such as T1 and ISDN. In a JPEG-only system, this chip is not required.

The AVP-1300-E performs single-chip MPEG encoding. The 1400-E version supports MPEG and Px64 compression and uses a fixed motion-estimation processor that searches exhaustively over a range of +/- 32 pixels with half-pixel accuracy. The bit rate is definable over a range of 40 Kbits/sec to 4 Mbits/sec (the latter is closer to the proposed MPEG-2 specifications) using a constant bit rate or a constant quantization step-size. Spatial resolution can go up to 720 pixels x 576 lines. The 1300-E can be used in the intraframe mode and thus can perform JPEG-like encoding.

The encoder contains several functional blocks. These include (a) a host bus interface, (b) FIFO to hold uncompressed data, (c) memory-controller to interface to
external DRAMs, (d) motion-estimator, (e) quantization processor to determine
quantizer step-size and output frame rate, (f) signal processor which is comprised of
six SIMD engines that perform DCT, quantization and zigzag scanning, (g) global
controller to sequence the operation of the blocks, (h) entropy coder to generate data
compliant with MPEG and H.261 standards, and (i) output FIFO to hold compressed data.

The decoder (AVP-1400-D) is less compute-intensive (primarily because it does not
need to perform any motion estimation). The AVP-1400-D supports decoding rates
of up to 4 Mbits/sec of compressed data and can support frame rates up to 30
frames/sec. Spatial resolutions up to 1024 lines x 1024 pixels can be supported (the
larger spatial resolutions are useful in JPEG still-image applications). The decoder
has an on-chip color space converter so that the decompressed YCrCb data can be
directly converted to RGB.

IIT Multistandard Chipset

The IIT (Integrated Information Technology) chipset [15] takes a different approach
than the AT&T chipset. It offers a programmable chip referred to as Vision Processor (VP). The VP has a 64-bit parallel architecture with a multiply/accumulate unit,
16 8-bit ALUs, and special circuitry for motion estimation. There are 25 and 33
MHz versions available. The microcode-based engine of this chip can execute
JPEG, MPEG and Px64 compression and decompression algorithms. The vision
processor does not perform MPEG encoding in real-time. JPEG, MPEG and Px64
decoding can be performed at 30 frames/sec at spatial resolutions of 352 x 240. For
MPEG and Px64 applications, the vision processor is combined with a vision controller (VC) chip which handles the pixel interface and frame buffer control functions. Recently, IIT announced the VCP which integrates most of the functions of
the VC and VP. The main functions within the VCP are depicted in Fig 5. This IC
can compress and decompress JPEG, MPEG and Px64. MPEG-2 decoding can be
performed in real-time whereas for encoding, several VCPs might be needed for
CCIR 601 resolution. VCP ICs will be available running at 40, 66, or 80 MHz.

NEC Multistandard Chipset

This chipset [16] is capable of real-time encoding and decoding of MPEG video.
Like the AT&T chipset it can support all three compression standards. Unlike the
AT&T chipset, the encoding and decoding algorithms are split across several chips.
The first chip performs inter-frame prediction, the second executes the DCT transform and quantization, and the third one does entropy-coding. These chips are
depicted in Fig. 6 (a-c). Motion estimation is performed in the prediction engine,
but unlike the AT&T's encoder, the motion search strategy is a two step process. In

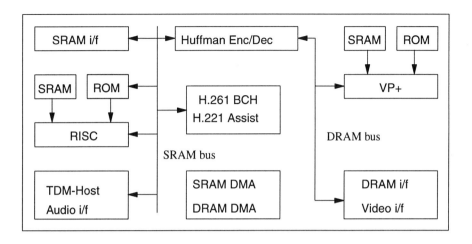

Fig. 5 : Block diagram of the IIT VCP processor.

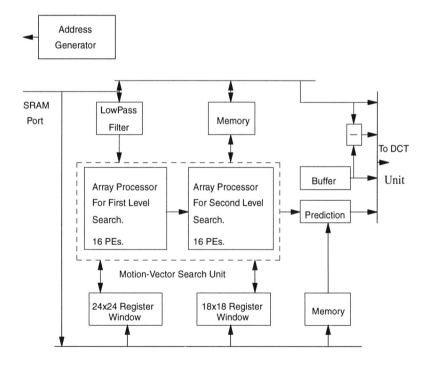

Fig. 6 a): NEC interframe predictor.

NEC's motion estimation algorithm, the first search step is over a 1/4 subsampled area and the second search step is performed over a 18x18 window and yields half-

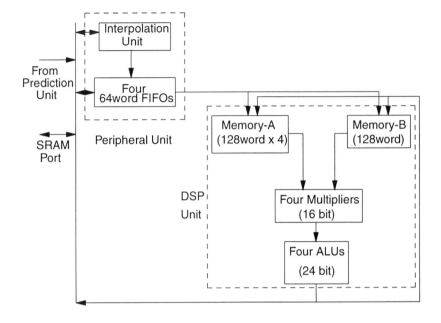

Fig. 6 b): NEC - DCT and Quantizer IC.

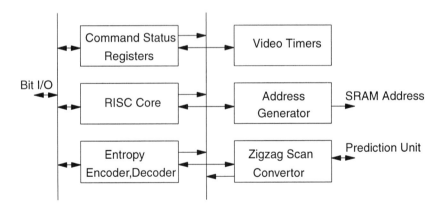

Fig. 6 c): NEC - Variable length coding IC.

pel accuracy. The DCT and quantization engine is a DSP unit which has four 16-bit MACs and 24-bit ALUs. The entropy coding engine has a RISC (Reduced

Instruction-Set Computer) core and performs all of the bitstream coding functions and also provides control for basic playback functions such as stop, play, and random access.

Other Vendors

Zoran[17] offers a two chip solution for use in still-image compression applications where the encoder bit rate needs to be a fixed rate. This is essential in applications such as digital still-video cameras. The Zoran chipset achieves the fixed rate via a two-pass process. During the first pass, the bit rate is estimated from the DCT values. The actual coding is performed during the second pass, by controlling the quantization process based on information derived by the first pass. Since much of the compression and decompression processes resemble DSP methods, traditional DSP IC vendors such as TI, Motorola, Analog Devices will probably announce JPEG and MPEG solutions based on an optimized DSP core. This trend was evidenced recently in a disclosure from Matsushita [18], 5 wherein, the 2 GOPs optimized DSP engine has been estimated to provide 15 frames/sec decoding of Px64 compressed video. A more detailed description of the Matsushita VDSP (vector digital signal processor) will be given in the next section. From the graphics side, IC vendors such as Brooktree and Inmos will have compression and decompression chipsets which will be based on an optimization of the graphics core in their current offerings. For acceptance within a multimedia computer, we believe that a chipset with a high-level of integration and support for host bus interface and DRAM control logic would probably be preferred.

4. General Purpose Image Processor ICs

Designs for general-purpose image processors (IPs) need to combine the flexibility and cost effectiveness of a programmable processor with the high performance of an ASIC. Therefore, it is no surprise that one main class of programmable image processors are extensions of general-purpose DSP chips. We call them *uni-processor* IPs. They include one multiply/accumulate (MAC) unit, separate buses for instruction and data, and independent address generation units. Depending on the target applications, they also include special arithmetic function units (for example, dual arithmetic units for the computation of absolute values or L2 norms in one cycle), on-chip DMA, and special data addressing capabilities. Another class of image processors, have an internal multi-processing architecture. We call them *multi-processor* IPs. Almost all these chips are developed for in-house use, are targeted to a specific range of applications (video processing, machine vision, data compression, etc.), and they don't seem to be commercially-available. Next, we will briefly describe ICs in both of these categories.

4.1 Uni-processor IPs

VISP

Fig. 7 shows a block diagram of the Video Image Signal Processor (VISP) from NEC [19]. This design is a very good representative of most programmable, uni-processor IPs that can be considered extensions of general-purpose DSPs. This is 16-bit video processor developed for real-time picture encoding in TV conference systems. It has two input and one output data buses, two local RAMs, (128 x 16-bit each in the DMU unit), an instruction sequencer (SCU), a 16-bit ALU, a 16 x 16-bit multiplier, a 20-bit accumulator, a timing control unit (TCU), an address generation unit (AGU), and a host interface unit (HIU). For efficient I/O, the processor has a built-in DMA controller in the HIU unit. The address generation unit can support up to three external buffer memories. Like **Hitachi's IP** [20], and the Digital Image Signal Processor (**DISP**) processor from Mitsubishi [21] it has hardware support to translate a two-dimensional logical address to a one-dimensional physical memory address.

VISP's ALU is specially designed to compute pattern matching operations in a pipelined manner. It can compute either L1 or L2 norms for vector quantization and motion compensation operations. Fig. 8 shows a block diagram of the ALU. The first stage has two shifters (SFT) for scaling the input video data. The second to fourth stages are for the ALU and the multiplier. The output of the multiplier is extented to 20 bits by the EXPAND module in the fifth stage. Data is further scaled by a 16-bit barrel shifter (BSF). The last stage is a minimum value detector (MMD) that is used for efficient motion picture coding.

In similar designs, **Toshiba's IP** [22] has dual arithmetic units to compute an absolute value in one cycle. For efficient I/O, the processor can be either in execution or data transfer mode. In the data transfer mode, the processor is controlled by a special data control unit which includes a 32-bit address register for accessing internal data memories or registers.

Hitachi's IP has special hardware for searching for minimum and maximum values 20. Furthermore, a special bit operations unit performs word-by-word bit manipulations. Using this unit, a pattern matching operation can be completed in only two steps. The chip achieves additional high I/O performance by using a two-level instruction hierarchy. It uses single-cycle "micro" instructions for internal data processing and two-cycle "pico" instructions for I/O with external data. The micro instructions activate the pico instructions, and data processing and I/O can proceed independently.

The Real-time Image Signal Processor (**RISP**) processor from Matsushita [23] has a multiplier/divider unit and a special flag counter for high speed averaging operations. Division uses the multiplier and an 11 x 256 reciprocal ROM.

Instead of on-chip DMA and data memories (as in Hitachi's IP, VISP and DISP), Toshiba's IP has hardware support for direct access to three external memories, and

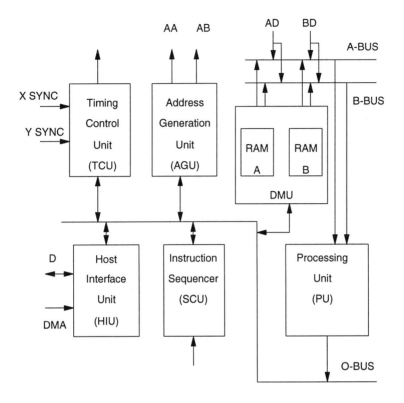

Fig. 7: Block diagram of VISP.

RISP has an array of 25 local image registers and buffers with a 25 to 1 selector. This feature, allows RISP to freeze and operate on a 5 x 5 array of image pixels at a time, and data addressing is performed by simply setting the multiplexors select code.

The VISP processor has a 25 ns cycle time, and like most of the others image processors can be used in a multiprocessor array.

VDSP

A processor of great interest is the Video Digital Signal Processor (VDSP) from Matsushita 5. This is a programmable IP, but with special enhancements in the architecture and the instruction set for more efficient implementation of coding standards, such as H.261, JPEG, and MPEG. Fig. 9 shows a block diagram of VDSP. It has three 16-b x 1024 data RAMs (M1, M2, and DM), each with its own address generator, an enhanced 16-b ALU, a 16 x 16-b multiplier (MPY), a DCT/IDCT circuit, a loop filter circuit (FLT), serial and parallel I/O control ports, and two 16-b

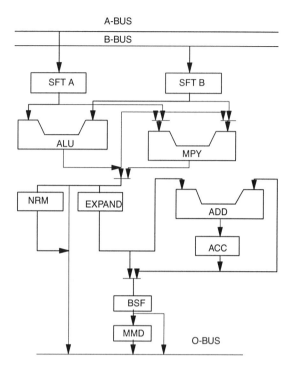

Fig. 8 : Block diagram of the VISP processing unit.

data ports (D1 and D2) with their own 24-b address generator units. The data RAMs can store a 32 x 32-pixel window. There is also a 1K x 32-b instruction RAM (IRAM) and a 32-b x 512 instruction ROM (IROM).

One of the main innovations in this IP is support for *vector pipeline* instructions. These instructions allow vector operations to be performed by a specified combination of the ALU, the multiplier, the DCT, and the loop filter. For example, a single instruction can execute commands like $Y = \sum |A_i - B_i|$, where A_i and B_i are data vectors. Vector-like operations are possible because of a two-dimensional space address generator (shown in Fig. 2). Given five parameters (start address, number of data per row, etc), an address sequencer generates automatically the full sequence of addresses needed for a particular vector operation. In addition to the conventional operations, the enhanced ALU performs also *min* and *max* of input data, clipping, and can compute the absolute value of the difference of two input data in one cycle.

Another feature of VDSP is an on-chip DCT/IDCT unit, and a loop-filter unit (as specified by H.261). The DCT core has eight ROM-accumulator units operating in

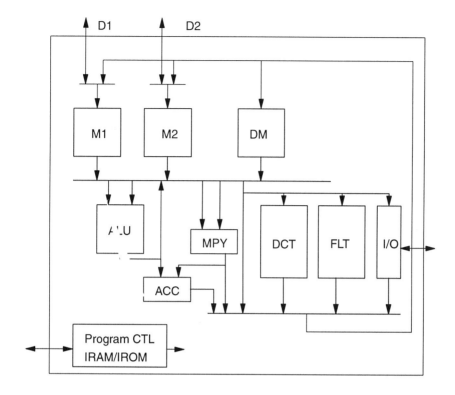

Fig. 9 : Block diagram of VDSP.

parallel and requires no multipliers. This IC runs at 60 MHz. An H.261 encoder can be implemented using two of these chips. The H.261 decoder implementation requires only one.

Most of the IP designs use fixed-point (16-b to 24-b) arithmetic. Recently, H. Fujii et al. [24] from Toshiba presented a floating-point cell library and a 100-MFLOPS (million floating-point operations per second) image processor (**FISP**). The floating-point IP has one multiplier, two ALUs, three data ports with address generation units, a data converter and local registers. The processor uses the IEEE754 floating-point format, but can also operate using 24-b fixed point arithmetic. The processor runs at 33 MHz.

4.2 Multi-processor IPs

The latest designs on image processing ICs combine features from both DSP architectures and classical SIMD and MIMD designs. Advances in integration allow now multiple processing units on a single chip. Thus, these designs combine the power of parallel architectures with the advantages of monolithic designs.

ViTec's PIP

Fig. 10 shows a block diagram of ViTec's Parallel Image Processor (PIP) chip [25]. This chip makes the core of ViTec's multiprocessor imaging systems [26]. It is probably the first image processor chip that employs an on-chip parallel architecture. Each chip has eight 8-bit ALUs and eight 8 x 8-bit parallel multipliers. A 9-way 32-bit adder is used to combine the results out of the multipliers. The architecture is highly efficient for convolution operations. A special replicator is used for replicated zooming. This architecture has limited local control, but it can provide up to 114 Million operations per second (MOPS), assuming an 8-bit wide datapath. Communication with an Image Memory (IMEM) is done via a 64-bit bidirectional bus. A separate 32-bit bus allows inter-processor communication in multi-chip systems. As in Hitachi's IP 20, the chip uses two levels of internal control. A 4-bit instruction, combined with a 6-bit address, selects a 64-bit microinstruction contained in an on-chip Writable Control Store.

The **Pipeline IP** from Matsushita [27] may be considered a variation of the ViTec design. It consists of nine processing elements positioned side by side. However, in this design the neighbor interconnections between the processing elements are reconfigurable. Each subprocessor consists of a 13 x 16-bit multiplier, a 24-bit accumulator, pipeline registers and multiplexors. The multiplexors allow different interconnection topologies among the processing elements, and image processing operations are computed in a systolic-type manner. The processor has three input and one output port, and several pipeline shift registers. The architecture is very efficient for matrix-type and filtering operations, and can compute an 8-tap DCT using a brute-force technique (matrix-vector multiplication).

ISMP

Fig. 11 shows the block diagram of the Image Signal Multiprocessor (ISMP)[28] from Matsushita. The ISMP is a multiprocessor version of the RISP 23 architecture that was referenced earlier. It has a main controller and arithmetic unit and four 12-bit processor elements (PEs). Each PE has a local image register, a 24-bit ALU and a 13x12-bit multiplier. It also has its own instruction RAM and local controller. The ISMP can be used to either operate at different locations of the image, or on the same location, but with four different feature extraction programs. ISMP chips can also operate in a multi-processor array.

To resolve the problem of feeding four different processors with data, the ISMP has a global image register and four local image registers. The global register consists of five 12-bit wide shift registers, each with a 12-bit input port, and drives an image bus of 300 lines. The local image registers, located in each PE, receive from the image bus a 5 x 5 image window. This window is latched at the falling edge of a special "start" signal, that starts the execution of each PE. The processor has a 20 ns

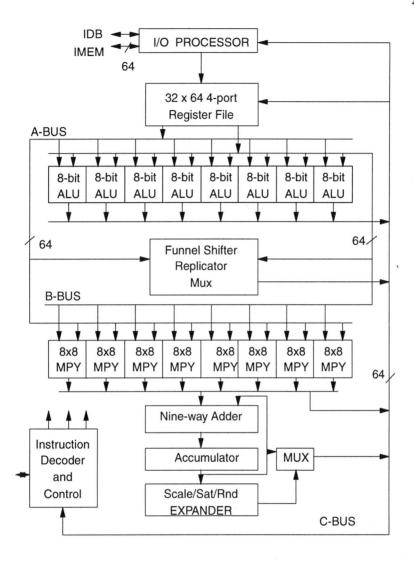

Fig. 10 : ViTec's Pipelined Image Processor

cycle time and can compute a 32-levels histogram of a 256 x 256 image in 23.6 ms.

IDSP

Fig. 12 shows another parallel architecture of a video signal processor from NTT, the Image Digital Signal Processor (IDSP) [29]. The processor has three parallel I/O ports and four pipelined processing units. Each processing unit has a 16-bit ALU, a 16 x 16-bit multiplier, and a 24-bit adder/subtracter. Each I/O port has 20-bit

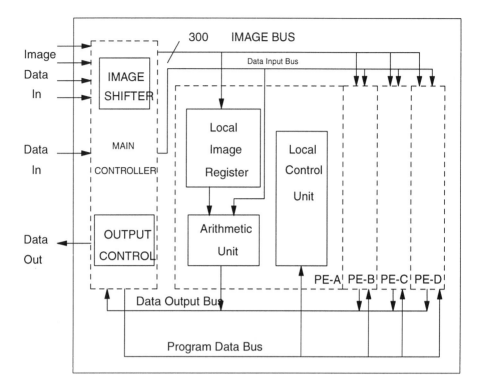

Fig. 11 : Block diagram of ISMP.

address generation unit, DMA control processor, and I/O control circuitry. Data transfers and program execution are executed in parallel.

Each Data processing Unit (DPU) has a local data cache memory and also shares a work memory. All memories are configured as 512-word x 16-bit dual-port RAMs. Data are transferred via ten 16-bit vector data buses and a 16-bit scalar data bus (SBUS). As in other architectures, where conventional VLIW (very large instruction word) would require hundreds of bits, the IDSP uses two levels of instruction decoding. IDSP has a 40 ns instruction cycle time.

TI's Video Processor (MVP)

Texas Instruments is working on an advanced processing chip aimed at image, graphics, video, and audio processing [30], [31]. It will be capable of supporting all the major imaging and graphics standards, such as JPEG, MPEG, PEX, and MS-Windows. It will have both floating-point and pixel processing capabilities for 2-D image processing and 3-D graphics.

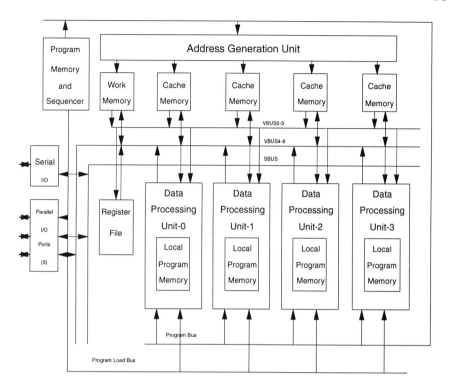

Fig. 12 : Block diagram of IDSP.

The chip will have multiple processors on a single chip (maximum eight). One of these will be a RISC type floating-point processor. The remaining processors will have a mixture of DSP and pixel manipulation capabilities. The on-chip processors are fully programmable and have instruction caches, thus each of the processors can run independently in a MIMD fashion. The processors and on-chip main memory are inter-connected via a crossbar. Performance estimates from TI are in the range of 3 billion "RISC-like" operations per second.

Table 5 summarizes the major features of the image processor ICs. Some notation issues: MPY and MAC stand for multiply and multiply/accumulate units, and BS denotes a barrel shifter. On the bit field, a dual number such as A/B denotes that A bits are used for input, but output or internal results may also be available with B bits of precision.

Table 6 shows benchmarking results for two classes of algorithms for which data for most of the above ICs is available: (a) a block type algorithm, such as the FFT or DCT, and (b) a spatial filtering algorithm with a 3 x 3 kernel.

Table 5
Programmable Image Processing ICs

System	Tech. (μm)	Clock (ns)	Instruction RAM	Data RAM	Processing units	bits
Toshiba	1.2 CMOS	50	64x16 16x99	None	ALU 2 AUs, BS 32x32 MPY	32
Hitachi	1.3 CMOS	50	1024x48 64x16	4x512x16	ALU BPU 16x16 MPY	16/32
RISP	bipolar	20	32x128	8x32 8x5x5	15x8 MAC/ Divider	8/15
ViTec's PIP	2.0 CMOS	70	64x64 64x16	32x64	8 8x8 MPYs 8 Adders	8/32
Pipeline IP	1.2 CMOS	30	None	None	9 13x16 MACs	13/24
VISP	1.2 CMOS	25	512x32	128x16	ALU, MAC Normalizer	16
DISP	1.0 CMOS	50	512x48	512x24x2 64x24- User Stack	ALU MPY BS	24
ISMP	1.2 CMOS	20	40x128/IP	12x25/IP	4 RISP IPs	12
IDSP	0.8 BiCMOS	40	512x32	5x512x16	4 DPUs	16/24
VDSP	0.8 CMOS	16.7	1.5Kx32	3x1Kx16	Enh. ALU MAC, 4xBS DCT/IDCT Loop FLT	16/24
FISP	1.2 CMOS	30	128x256	None	2 ALUs MPY	24/32 float

5. Discussion and Conclusions

As noted before, image processing architectures range in complexity from massively parallel systems to single chip ICs. However, regardless of their complexity, all architectures, from the multiprocessor to the monolithic, attempt to provide solutions for a common set of image processing issues. Table 7 shows a set of common problems in image processing and how they are being addressed by different architectures. Apparently, there is no unique or "correct" solution, since every architecture is optimized for a certain class of applications. For the designer of an image

Table 6
Benchmarking Results
for Programmable Image Processing ICs.

Processor	Transform	Filtering (256 x 256 data, 3 x 3 mask) ms
Toshiba	FFT, 1024 points 1.0 ms.	29.5
Hitachi	FFT, 512 points 1.5 ms.	39.3
RISP		11.8
DISP	FFT, 512 points 0.91 ms.	29.6
VISP	2-D DCT, 256x256 26.3 ms.	14.8
Pipeline IP		1.98
ISMP		3.9
IDSP	2-D DCT, 8x8 41.8 μs	
VDSP	2-D DCT, 8x8 3.0 μs	

processing IC, the design constraints are even bigger due to the limitations in silicon area and the wide range of requirements in image processing algorithms.

Looking back into the evolution of the present DSP designs, after the TMS32010 a new generation of DSPs emerged from manufacturers such as: AT&T, AMD, Thomson, Motorola, and TI. Each design had its own unique features, but all of them shared a common set of basic characteristics. A *generic* DSP design that combines those characteristics has a modified Harvard architecture with separate data and instruction buses, a multiply/ accumulate unit, at least two data memories with their own address generation units and bit-reverse circuitry, instruction memory, a microsequencer with a VLIW instruction set, and serial/parallel ports for real-time DSP applications. Can one define a similar *generic* architecture for programmable image processors? The latest designs from TI, Matsushita, and NTT point indeed to a generic design that combines features from both general purpose DSPs and special purpose image processors.

Fig. 13 shows the block diagram of such a design. It features a parallel architecture with multiple image processing units (IPUs), independent I/O and program execution, and support for efficient multi-dimensional addressing and multi-processor

Table 7
Common Issues in Image Processing Architectures

Problem	Solution
Processing Power	Special ASICs. Allow for multi-processing. Custom on-chip ALUs. On-chip multiprocessing.
2-D Addressing	Independent address generation units. Multiple data memories. Integrated, block-type addressing capability.
Data storage and I/O	Independent I/O and program execution. Multiple buses with interconnection routers. On-chip DMA.
Instruction sequencing and control.	On-chip microsequencers. Multi-level instruction sequencing. Host-dependent operation.
Precision.	Application dependent. From 8-bit to floating-point architectures.

configurations. Each IPU consists of a multiply/accumulate unit with local data cache and program control, or it is an application specific unit (i.e., a DCT processor for image compression, or a dedicated histogram generation unit.) The number of the IPUs depends on the target application and the available silicon area.

In the compression area, as the video compression standards reach their final form, image compression chips will be developed for applications in video-teleconferencing and high-definition TV (HDTV). These chips will probably include support for interfacing to digital networks like FDDI, will possess compute power in the range of 2-4 Giga operations per second for HDTV applications, and will include DRAM control for glueless interfacing to display memory. To facilitate the integration of audio applications, these ICs will also include a conventional DSP core for audio processing.

In summary, image and video processing acceleration circuitry will be an integrated part of future personal computers and scientific workstations. In this paper, we reviewed recent architectures for image and video processing ICs. Application specific processors are available for image and video compression, or low-level image processing functions. Programmable image processing ICs are either modified extensions of general purpose DSPs or incorporate multiple processing units in an integrated parallel architecture. Because of the iterative nature of low-level imaging

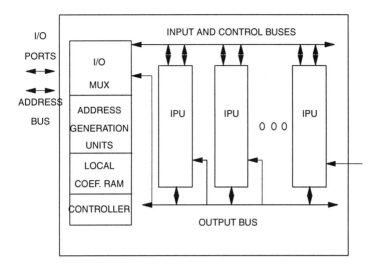

Fig. 13 : Block diagram of a generic image processor.

functions, such parallel architectures are ideal for high-speed image processing. A new generation of programmable image processors will soon emerge to accommodate the special needs of real-time video processing. They will consist of multiple computing units, some of them application specific, special 2-D data address generation units, and efficient I/O ports.

REFERENCES

1. M.D. Edwards, "A review of MIMD architectures for image processing," J. Kittler and M.J.B. Duff (eds.),*"Image processing system architectures,"* Research Studies Press, Letchwork, Hertfordshire, England, 1985, pp. 85-101.

2. S. Yalamanchili, K.V. Palem, L.S. Davis, A.J. Welch, and J.K. Aggarwal, "Image processing architectures: a taxonomy and survey," in *Progress in Pattern Recognition 2*, L.N. Kanal and A. Rosenfeld (eds.), Elsevier Science Publishers, North Holland, 1985, pp. 1-37.

3. T.J. Foundain, K.N. Matthews, and M.J.B. Duff, "The CLIP7A image processor," IEEE Trans. on Pattern Analysis and Machine Intelligence, Vol. 10, No. 3, pp. 310-319, May 1988.

4. R.D. Fellman, R.T. Kaneshiro, and K. Konstantinides, "Design and evaluation of an architecture for a digital signal processor for instrumentation applications," IEEE Trans. on ASSP, Vol. 38, No. 3, pp. 537-546, March 1990.

5. K. Aono, et al., "A video digital signal processor with a vector-pipeline architecture," IEEE J. of Solid-State Circuits, Vol. 27, No. 12, Dec. 1992, pp. 1886-1893.

6. K. Kitagaki, et al.,"A new address generation unit architecture for video signal processing," SPIE Vol. 1606 Visual Commun. and Image Processing '91: Image Processing, pp. 891-900, 1991.

7. M. Kidode and Y. Shiraogawa, "High speed image processor: TOSPIX-II," *Evaluation of Multicomputers,*" L.M. Uhr et al. (eds.) Academic Press, New York, 1986, pp. 319-335.

8. N.L. Seed, A.D. Houghton, M.J. Lander, and N.J. Goodenough, "An enhanced transputer module for real-time image processing," Third Intern. Conf. on Image processing and its applications, IEE, Hitchen, Herts, England, pp. 131-135, 1989.

9. K.S. Mills, G.K. Wong, and Y. Kim, " A high performance floating-point image computing workstation for medical applications," SPIE Vol. 1232, Medical Imaging IV: Image capture and display, pp. 246-256, 1990.

10. *Digital Signal Processing Databook*, LSI Logic Corporation, September 1991.

11. W.K. Pratt, *Digital Image Processing,* John Wiley and Sons, New York, 1991.

12. JPEG-1 DIS Working Group 10, Draft International Standard, DIS 10918-1, CCITT Rec. T.81, Jan. 2, 1992.

13. MPEG-1 CD, Working Group 11, Committee Draft ISO/IEC 11172, Intern. Standards Organization, IPSJ, Tokyo, Dec. 6, 1991.

14. Video Codec for Audiovisual Services at p x 64 Kbits/s. CCITT Recommendation H.261, CDM XV-R 37-E, International Telegraph and Telephone Consultive Committee (CCITT), Aug. 1990.

15. D. Bailey et al., "Programmable vision processor/controller for flexible implementation of current and future image compression standards," IEEE Micro, October 1992, pp. 33-39.

16. I. Tamitani et. al, "An encoder/decoder chip set for the MPEG video standard,", IEEE ICASSP-92, pp. V-661 - V-664, San Francisco, 1992.

17. A. Razavi, R. Adar, et. al., "VLSI implementation of an image compression algorithm with a new bit rate control capability", IEEE ICASSP-92, pp. V-669 - 672, San Francisco, 1992.

18. T. Araki et. al., "The architecture of a vector digital signal processor for video coding", IEEE ICASSP-92, pp. V-681 - V-684, San Francisco, 1992.

19. K. Kikuchi, et al., "A single chip 16-bit 25 ns real-time video/image signal processor," IEEE J. of Solid State Circuits, Vol. 24, No. 6, pp. 1662-1667, Dec. 1989.

20. K. Kaneko, et al., "A 50 ns DSP with parallel processing architecture," IEEE ISSCC 87, pp. 158-159, 1987.

21. T. Murakami, K. Kamizawa, M. Kameyama, and S. Nakagawa, "A DSP architectural design for low bit-rate motion video codec," IEEE Trans. on Circuits, and Systems, Vol. 36, No. 10, pp. 1267-1274, Oct. 1989.

22. A. Kanuma, et al., "A 20 MHz 32b pipelined CMOS image processor," IEEE ISSCC 86, pp. 102-103, 1986.

23. H. Yamada, K. Hasegawa, T. Mori, H. Sakai, and K. Aono, "A microprogrammable real-time image processor," IEEE Journal of Solid State Circ., Vol. 23, No. 1, pp. 216-223, 1988.

24. H. Fujii, et al., " A floating-point cell library and a 100-MFLOPS image signal processor," IEEE J. of Solid-State Circuits, Vol. 27, No. 7, July 1992, pp. 1080-1087.

25. J.P. Norsworthy, D.M. Pfeiffer, M.K. Corry, and J.A. Thompson, "A parallel image processing chip," IEEE ISSCC, pp. 158-159, 1988.

26. D. Pfeiffer, "Integrating image processing with standard workstation platforms," Computer Technology Review, pp. 103-107, Summer 1991.

27. K. Aono, M. Toyokura, and T. Araki, "A 30 ns (600 MOPS) image processor with a reconfigurable pipeline architecture," IEEE 1989 Custom Integr. Circuits Conf., pp. 24.4.1-24.4.4, 1989.

28. M. Maruyama et al., "An image signal multiprocessor on a single chip," IEEE J. on Solid-State Circ., Vol. 25, No. 6, pp. 1476-1483, Dec. 1990.

29. T. Minami, R. Kasai, H. Yamauchi, Y. Tashiro, J. Takahashi, and S. Date, "A 300-MOPS video signal processor with a parallel architecture," IEEE J. of Solid-State Circuits, Vol. 26, No. 12, pp. 1868-1875, Dec. 91.

30. R.J. Gove, "Architectures for single-chip image computing," SPIE Electronic Imag. Science and Techn. Conf. on Image Processing and Interchange," San Jose, Feb. 1992.

31. K. Guttag, R.J. Gove, and J.R. Van Aken, "A single chip multiprocessor for multimedia: The MVP," IEEE CG&A, Nov. 1992, pp. 53-64.

3

HIGH PERFORMANCE ARITHMETIC FOR DSP SYSTEMS

G.A. Jullien, Director
VLSI Research Group, University of Windsor
Windsor, Ontario, Canada N9B 3P4

INTRODUCTION

Digital Signal Processing, since its establishment as a discipline 30 years ago, has always received a great impetus from electronic technological advances. It often rides the crest of that wave and sometimes is responsible for pushing it.

General purpose DSP chips first appeared in the late 1970's, and were essentially microcontroller technology with most of the area given up to an array multiplier. The architectural technology was pushed by the requirement for large numbers of multiplications to be performed at high speeds. As technology improved, the multiplier shrank in relative area, to the point where more than one could be placed on the chip. The need to amortize development costs over large volumes of production chips, led DSP chip design in similar directions to that of single-chip general processor design; instruction based bus architectures with a handful of 'fast' floating point ALUs on board. The chips are easy to use, and DSP algorithms can be 'firm-wired' by OEM houses in a very short time. The raw processing power of the silicon, however, is sacrificed for the convenience of programmability over a wide variety of algorithms.

Special purpose DSP chips also saw their birth in the late 1970's. Special purpose architectures are programmable only in a local algorithmic sense (e.g. a FIR filter fixed architecture with programmable coefficients) but the potential processing speeds can be 1 to 2 orders of magnitude greater than those obtained by general purpose programmable designs. It is this latter category of chip architectures that this chapter addresses, and we will briefly cover a range of special arithmetic implementation strategies that can be used to advantage in situations where the chip is being designed for a particular purpose in mind.

It is important to define the widely used terms *speed* and *fast*. They are often used synonymously, but *fast* implies a measurement in time only, whereas *speed* connotes a rate. For DSP that rate is surely samples per second, and a high speed design should refer to a circuit that processes large numbers of samples per second. Unfortunately the term is not always used appropriately, and this can also lead to some basic confusion in the design decisions that are to be made in building either *fast* or high *speed* systems. As an example of this confusion, an engineer required to design a high speed multiplier, will probably seek out the *fastest* multiplier available, and then place latches at the input or output, or even within the multiplier, in order to create a data pipeline. In this case the computation is being carried out in a fast manner and only the data is being pipelined at high speed. This arrangement works, and is often used, but it is not necessarily the most efficient use of resources. For high speed designs, we need to look at pipelined computations, rather than pipelined data that is the result of fast computations. This issue is a theme that runs through the chapter, and we will review a recent circuit solution to generic pipelined computations.

It is probable that many DSP solutions require high speed, but there are also cases where fast solutions are preferred. For example, if we wish to build a digital filter to filter a continuous data stream, then a high speed architecture is desired; on the other hand, we may wish to compute an FFT algorithm within the lowest period of time possible, and this will require a fast solution. This chapter will also briefly review recent work at both the architectural and circuit level in the area of fast designs for arithmetic circuits.

Concurrent with the search for high performance arithmetic circuits, is the basic number representation used, and the associated algorithms to implement arithmetic operations; we will discuss some of the issues here, and will concentrate on systems that have fixed parameters where efficiencies can be obtained by pre-computation of results of arithmetic operations.

To summarize. This chapter is concerned with techniques for building high performance arithmetic (both *fast* and *high speed*) for special purpose DSP processors. We look at the topic from the areas of number representation, arithmetic algorithms, and circuits in order to allow selection of the most appropriate vehicle for a specific high performance DSP design.

SOME ISSUES WITH NUMBER REPRESENTATIONS

A complete study of number representations is considerably beyond the scope of this chapter; our goal is not to present a tutorial on the subject, there are many excellent books available [1][2][3], but to present a small selection of relevant issues that might normally be overlooked. A restriction that we will immediately impose is only to discuss fixed point (or integer) arithmetic. Although floating point arithmetic is widely used in current generations of general purpose DSP chips, high performance systems will almost always use fixed point arithmetic. We will also concentrate on basic operations of multiplication and addition, since these are by far the most used operations within many DSP algorithms; they are also able to be performed precisely within a finite precision range. Examples of multiply/add only architectures are FIR filters and inner product type transforms such as DFT and DCT. There are normally provisions for limiting the dynamic range growth of fixed point representations, and so we may also include IIR filters and the many fast transform architectures [4] that require many cascades of multiplications along a signal path.

The most popular approach to performing DSP arithmetic is to use some form of binary number representation; usually a choice between twos complement and sign-magnitude. There are many reasons for this, but the major one is probably familiarity. Performing arithmetic operations uses the same algorithms as in any other radix, particularly 10 with which we are all very familiar. There are structural niceties as well: complete algorithms can be performed with replications of the same adder blocks and basic translations of the algorithms lead to very regular arrays, though not necessarily the most efficient. There are many other forms of representation into which binary representations can be mapped, and we will discuss selected issues related to these other forms of number representation.

The Curse of the Carry

Arithmetic based on strict binary representations are always dominated by carry manipulations. Essentially, the carry contains important information about weight overflow, and is always present in any weighted magnitude representation, such as binary. Adding two binary numbers (we will drop the use of 'representation' when it is obvious by context) that have 1's in all digit positions will require transmission of the carry from the least significant bit (LSB) to the most significant bit (MSB) and will propagate to an invisible higher bit position. This we will define as the *curse of the carry*; it takes considerable time for this propagation to occur. If we examine any architecture for fast arithmetic, the manipulation of the carry is the defining objective. Examples are: Manchester Carry Chains; Carry-Look Ahead; Carry Skip; Carry Save etc.[3]. Our lowly basic adder uses Carry Ripple.

Many of the alternate number representations are studied because the carry manipulation is improved. Examples are *Redundant Arithmetic* [5], where the carry is still present, but does not ripple across the entire representation, and *Residue Number Systems* (RNS) [6], where the carry is removed completely, but at considerable expense in representation mapping.

Multiplicative Complexity

Multiplication is as common as addition in many DSP algorithms, but is considerably more expensive. If we consider multiplying positive numbers only, then multiplication of two numbers is polynomial multiplication of the individual bits of the representations, as shown in eqn. (1) for a binary representation with B bits for each number.

$$x \times y = \left(\sum_{i=0}^{B-1} 2^i x^{[i]} \right) \left(\sum_{j=0}^{B-1} 2^j y^{[j]} \right) \tag{1}$$

As all signal processing engineers know, multiplying two polynomials is the same as convolving their coefficient sequences, and so if we consider multiplying $(2^B - 1)(2^B - 1)$ (both sequences are all ones) then we have the analogous situation of convolving two rectangular functions with a resulting triangular output as seen in Figure 1 (for $B = 8$); the dots represent binary 1's, the result of multiplying 2 adjacent 1's together in the convolution operation (this multiplier is a simple AND gate). The columns of AND plane bits are compressed into a double length row of single bits using compressors. The simplest compressors are full and half adders. Figure 2 shows a time optimal array of full and half adders (the half adders are shown shaded) for $B = 8$. The long rectangle at the bottom is a fast adder for the final row compression. The array follows a sequence of bit compression that Dadda showed to be optimum in terms of critical path [7].

The obvious doubling of the output number range does not really impact on a strict binary implementation, since we are only concerned with placing bit compressors to reduce the bit columns to a single bit. We do not have to use up hardware to hold the upper 0's of the two numbers. For non-weighted representations, however, we will require to set up sufficient dynamic range for each of the input numbers to handle this increase.

Figure 1. Multiplication Viewed as Bit Sequence Convolution

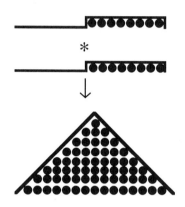

Figure 2. Dadda's Fast Column Compressor

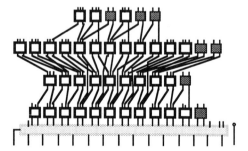

In effect we have introduced redundancy into the input number representation in order to improve the arithmetic hardware. For a standard binary implementation, there is always a final adder required to reduce the multiplier output to a single row of bits and for a fast multiplier this will have to be constructed carefully in order to avoid its domination of the critical path.

For high speed multipliers, a regular array architecture is normally chosen [8]. This corresponds to a direct addition of partial products; the partial product array is shown in Figure 3. Also shown is the full/half adder array that is used to sum the partial products. Note that both the array of AND plane outputs and the adder array, contain the same minimum number of bits and adders as the arrays of Figure 1 and Figure 2. It is clear that the array of Figure 2 has a lower critical path, but that the array of Figure 3 is much more regular (this latter architecture is often referred to as an *array multiplier*). The regularity of the array multiplier can easily be converted to a pipelined high speed array by placing lines of pipeline latches between each row of adders. Depending on the throughput rate required we can place the latches less frequently if desired.

Figure 3. Array Multiplier

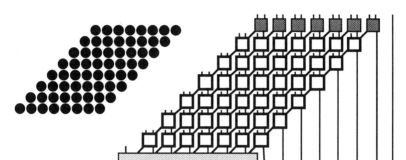

Reducing Multiplicative Complexity

If we examine the AND plane inputs to the array multiplier, we find that the nth row has a common AND plane input which is the nth bit of the multiplicand (or multiplier). If we have *a priori* knowledge of this bit pattern, as, for example, with fixed coefficient filters, then we may leave out a complete row from the array. If coefficients are selected that have minimal numbers of 1's in their bit pattern, then this technique can be quite efficient. It has an unfortunate side-effect, however, in that certain input sequences may excite more multipliers with large numbers of rows, thus causing an increase in power consumption. Filters whose power requirements changes with the type of input sequence are not really commercially viable.

We may also invoke a technique first discovered by A.D. Booth in the early years of computer development [9]. The basic method is covered in almost every book on digital logic, so we will not discuss it here, except to point out that current techniques apply it to a mapping, or recoding, technique [10] in order to reduce the number of stages required. Even so, this reduction is not as large as the fixed coefficient simplifications, on average.

For those of us brought up on logarithmic tables and slide-rules, multiplicative complexity was reduced to addition, providing we did not object to looking up, or mapping, between the logarithmic and linear representations. The slide-rule stored the tables as graduations on a set of scales; the act of moving the slide performed the required logarithmic addition. Multiplication was always a much easier operation than addition since addition had to be performed using pencil, and paper. We may use the same mapping principles, electronically, to reduce multiplicative complexity; this is known, generically, as the *Logarithmic Number System* (LNS) [11]. This has been proposed for use in DSP systems for many years, but, interestingly, often for the dynamic range increase rather than the reduction of multiplicative complexity [12]. The mapping overhead can be reduced if a chain of calculations are able to be performed within the mapped system; this normally implies being able to perform addition within the logarithmic system, a difficult operation, so now additive complexity is the issue! In any mapping scheme, if a number to be mapped is a constant then the mapped value of the constant can be stored rather than the original constant. This can be used with the LNS to further remove multiplicative complexity. There are other issues to be considered regarding the LNS, and these relate to imprecision and non-linearity. These two issues are related, since the non-linearity will not be important unless there is imprecision in the calculations. This will be briefly discussed in the next section.

The RNS [6] is an exotic technique, that requires extensive mapping circuitry, but can produce remarkable reduction in multiplicative complexity if used appropriately. The method is based on isomorphic mapping procedures between a single finite ring and a direct product of rings. We define a ring of integers as in eqn. (2).

$$\Re(M) = \{S : \oplus_M, \otimes_M\}; S = \{0, 1, ..., M-1\} \tag{2}$$

where $a \oplus_M b = (a+b) \operatorname{Mod} M$ and $a \otimes_M b = (a \times b) \operatorname{Mod} M$. If we make M large enough, then we can avoid the modulo *wrap-around* and emulate ordinary arithmetic. We go to these lengths so that we can invoke the direct product ring mapping feature. Let $a_i = a \operatorname{Mod} m_i$ etc. where $\{m_i\}$ are a set of relatively prime numbers and $M = \prod_{i=1}^{L} m_i$. Then if we map 2 elements of $\Re(M)$, say a and b, by their residues $\{a_i\}$ and $\{b_i\}$ then the following remarkable properties exist, as given in eqn. (3).

$$a \oplus_M b \Leftrightarrow \{a_i \oplus_{m_i} b_i\}$$
$$a \otimes_M b \Leftrightarrow \{a_i \otimes_{m_i} b_i\} \tag{3}$$

Thus, direct products rings allow independent (and therefore parallel) multiplications and additions over relatively small rings (5 or 6-bits usually) and all of the ring computations can be pre-stored in tables of no more than 12 address bits. Computing with constants is particularly easy since the constants can be included in the tables at essentially no extra cost [13]. Note, with the exclusive use of tables for computation, multiplicative complexity and additive complexity are identical. The major problem with this technique is the reverse mapping; we use the Chinese Remainder Theorem (or some equivalent form) as shown in eqn. (4).

$$X = \sum_{k=1}^{L} {}_M \{\hat{m}_k \otimes_M [x_k \otimes_{m_k} (\hat{m}_k)^{-1}]\} \tag{4}$$

with $\hat{m}_k = \dfrac{M}{m_k}$, $X \in R(M)$, $x_k \in R(m_k)$ and $(x)^{-1}$ the multiplicative inverse operator. We have also used the notation $\sum_M ...$ to indicate summation over the ring, $R(M)$.

Because of the mapping overhead, RNS and LNS architectures are more appropriate for high speed pipelined systems. We will show a later case study, where the complexities of the CRT can be implemented using high speed pipelines.

Combining RNS mapping with finite field logarithms is particularly impressive. The finite field logarithms are closed over the field, yielding exact results while still retaining additive complexity of multiplication. This technique has lead to a successful commercial implementation of convolution using RNS arithmetic [14].

The essence of the finite field logarithm technique (or index calculus, as it is sometimes called) is related to the isomorphism that exists between additive and multiplicative groups of a finite field. The isomorphism is defined in eqn. (5).

$$g_n \otimes_{m_i} g_j \Leftrightarrow \rho^{\left(k_n \oplus_{m_i - 1} k_j \right)} \tag{5}$$

ρ is a generator of the field GF (m_i) (a Galois Field) where m_i is one of a set of prime moduli. The multiplicative group elements are entirely generated from powers of ρ, and so we may use ρ as a base for our logarithmic technique. Since we will choose fields with a reasonably small number of elements, the equivalent log table look-ups can be stored very efficiently, and precisely, in memory. This technique has been used for many years to reduce multiplicative complexity in RNS DSP systems [15][16][17][18]. If the field moduli are small, then the mapping overhead can be easily accommodated within the index addition circuitry. For larger fields, an interesting technique that performs addition within the log (or index) system has been recently published [19].

Many other methods exist for minimizing multiplicative complexity. We should not leave this subject without mentioning the *Quarter Squares* technique. This method has been known, probably, for centuries; it seems to have been first mechanized in the construction of multipliers for analog computers. The technique is embodied in the relationship of eqn. (6).

$$x \times y = \frac{1}{4} \left((x+y)^2 - (x-y)^2 \right) \tag{6}$$

The difficult binary (two-variable) operations are addition and subtraction, the other operation required is unary, a squaring mechanism. Since squaring is much easier than general multiplication, we have reduced the asymptotic complexity of multiplication. We have replaced it, however, with a rather more complex data flow problem. This technique probably works best over small dynamic ranges, where squaring can be reduced to look-up tables, and is particularly suitable for RNS mapped systems

Imprecision

The issues of imprecision in both the parameters and arithmetic processing of DSP systems has been well explored, mainly because of hardware limitations in the early years of DSP. For designers using general purpose DSP chips, with floating point arithmetic units, imprecision is not a major issue, but for systems that require high performance, fixed point solutions are the only consideration. The imprecision of DSP parameters and arithmetic operations are somewhat separate issues.

Parameter imprecision perturbs the function of the DSP algorithm. For example, a filter designed to a specific frequency response, will have a slightly different response if the precision of the parameters is reduced. To take this into account, the filter design algorithm will optimize the parameters based on the precision desired. For fixed coefficient reduction of multiplicative complexity, using standard binary arithmetic, it is important that parameters also be sought that have minimum numbers of 1's in their representation. For other weighted representations, it is simply enough to limit the coefficient dynamic range. A non-optimum, but simple technique, is to round the parameters within the dynamic range chosen. As an example, we can consider the effect of rounding the coefficients of a FIR filter to a prescribed dynamic range. Let the desired transfer function be given by eqn. (7).

$$H(z) = \sum_{i=0}^{N-1} h_i z^{-i} \tag{7}$$

The $\{h_i\}$ are assumed to be precisely represented. The effect of imprecision is to implement the filter with a different set of coefficients, $\{\tilde{h}_i\}$, where $\tilde{h}_i = h_i - \varepsilon_i$, the ith coefficient, is representable within the precision limits imposed by the hardware, and ε_i is the error of the ith coefficient, assumed bounded by $\frac{q}{2}$, where q is the smallest representable number above zero (the quantization step). This yields an error transfer function, from eqn. (7), given by eqn. (8).

$$E(z) = \sum_{i=0}^{N-1} e_i z^{-n} \tag{8}$$

The effect on the frequency response can be found by evaluating the error transfer function on the unit circle, yielding eqn. (9).

$$|E(e^{j\omega t})| \le \sum_{i=0}^{N-1} |e_i| \le \frac{Nq}{2} \tag{9}$$

This gives an upper bound on the error of the frequency response, and can be used to determine the allowable precision reduction parameterized by q.

The affect of arithmetic imprecision is to inject digital noise into any node that employs some form of precision reduction. The most usual scenario is to employ bit length reduction (scaling), with rounding, after a multiplication operation. In operating DSP systems with large dynamic range and wide bandwidth signals, the noise injection due to scaling operations within the architecture, can be considered to be zero mean, uniformly distributed random variables, uncorrelated with each other [20]. The distribution of the variables is over the range $\left(-\frac{q_s}{2}, \frac{q_s}{2}\right)$, where q_s is the quantization step associated with the scaling operation. The effect of a noise source on the output signal can be found from the transfer function between the source and output. The situation is shown in Figure 4. The overall transfer function is assumed to be $\frac{Y(z)}{X(z)} = H(z)$, and noise is injected at the ith injection point in the structure where the transfer function to the output is $H'_i(z)$. The steady-state noise variance at the output is given by eqn. (10) [20].

$$\sigma^2_{n_i} = \frac{q^2_s}{12} \cdot \sum^{\infty}_{j=0} h'^2_{i,j} = \frac{q^2_s}{12} \cdot \frac{1}{2\pi j} \oint H'_i(z) H'_i\left(\frac{1}{z}\right)\frac{1}{z}dz \qquad (10)$$

Figure 4. Injection of Digital Noise

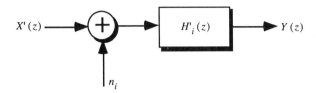

If we assume that the signal at the output is also a zero mean random variable, uniformly distributed across the entire output dynamic range of B bits, then we have the rather optimistic signal to noise ratio given by eqn. (11).

$$\frac{\sigma^2_y}{\sigma^2_{n_i}} = \frac{B^2}{\frac{1}{2\pi j}\sum^{S-1}_{i=0} \oint H'_i(z) H'_i\left(\frac{1}{z}\right)\frac{1}{z}dz} \qquad (11)$$

where S is the number of noise injection sources. This S/N ration is optimistic, because we cannot guarantee that the output will 'fill' the output dynamic range of B bits, let alone look like a uniformly distributed random variable. We can, however, scale the transfer function between signal and output so that the dynamic range is usefully used. For example, borrowing the idea of the signal as a uniformly distributed random variable over B bits at the input, we can insert a scaling factor, K, as evaluated in eqn. (12), so that the output has the same variance as the input.

$$K^2 = \frac{1}{\frac{1}{2\pi j}\oint H(z) H\left(\frac{1}{z}\right)\frac{1}{z}dz} \qquad (12)$$

This scaling factor is optimistic, in the sense that the input and output signal energies are the same; there is no guarantee that the dynamic range will not overflow for some output samples. Practice shows, however, that this is a reasonable approach to take in maximizing utilization of a given dynamic range.

All of these discussions assume that we will be using fixed point arithmetic, with a linear number representation. LNS arithmetic is more difficult to analyze in terms of noise injection (because of the non-linearity), though the overall dynamic range is larger than a linear range for a given number of bits. If we use floating point arithmetic, then for a mantissa size equal to the fixed point word width, the floating point realization will be superior. If we include the exponent within the number of bits (a more realistic comparison) then conditions can be found where either fixed or floating point is better [21]. For an RNS realization, scaling is a major expense and so we will normally define a dynamic range equal to the maximum computational range within the DSP structure; all signals, including the input, will have representations over this range. This represents a degree of redundancy not experienced in a standard binary representation. Because scaling is expensive, we will undoubtedly arrange the architecture to

68

limit scaling; this will translate to increased signal to noise ratios at the output. For the case of occasional overflow samples, the binary implementation can allow simple circuit extensions to saturate the arithmetic on overflow occurrence (any overflow simply maintains a constant output at the maximum value). For RNS implementations, overflow can be more severe since overflow detection is expensive, and often avoided. In this case, the output after overflow is the residue of the true result, modulo M, and this residue can take on any value within this dynamic range. This can be quite catastrophic for systems with feedback (e.g. recursive filters) and severe overflow oscillations can result. For this reason, the most useful algorithms to be implemented with RNS are entirely feed-forward.

Direct Product Rings

Mapping a DSP calculation into a direct product ring has many architectural and arithmetic advantages. It also has some considerable arithmetic disadvantages which are often quoted as reasons for not using the mapping. Under certain circumstances, however, these mappings are to be encouraged. One mechanism for mapping to direct product rings has already been introduced; the Residue Number System (RNS). There are other procedures available which will be discussed later. Interestingly enough, the use of RNS is normally recommended because of the carry-free nature of the computations; the structural advantages, however, should not be overlooked, and, in fact, may be the sole motivation for using the mapping when considering VLSI implementations.

The carry-free structure arises from the complete independence of the direct product ring calculations, but this is also the reason why the structure is so amenable to VLSI implementation. Often, simple word-level DSP architectures hide the more complicated structure (and data flow) associated with the bit-level implementation. As an example, consider the apparently very simple systolic array for convolution, as shown in Figure 5.

Figure 5. Systolic Array for Convolution

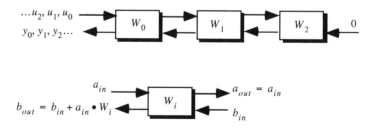

The array is systolic, so the basic multiplier/accumulator is implemented in a pipeline fashion. If we are to implement a general multiplier, then an array, such as that of Figure 3 will be required together with an adder. The array is able to be pipelined at the full adder level (each row) requiring a number of pipeline stages proportional to the word width of the multiplicands used as coefficients. If we look at the bit-level data flow, then it is two dimensional, with data moving both downwards and towards the MSB end of each adder row. This two-dimensional data flow can cause clock distribution problems for systems that are pushing the implementation technology for as high a throughput rate as possible. Testing large chips is also a function of the two-dimensional connectivity, though regular arithmetic arrays are often C-testable; that is, they can be tested with a small number of test vectors scanned through the entire array. Thus we see that the apparently elegant linear systolic array has hidden complexities at the bit level.

An implementation using direct product rings will simply have a replication of the linear systolic structure for each ring. If the rings are small (i.e. a ring modulus, say, ≤ 32 or 5-bits) then we replicate the linear array for each ring, independently. This produces a structure with fewer pipeline stages and improved clocking distribution. We can provide, for example, separate clock lines for each ring array, without the necessity of synchronizing latches at the LSB and MSB ends of the computational range. We have to accept the complexities of mapping into and out of the RNS and also the problems of implementing modulo operations; this is where the extensive use of look-up tables enters the picture (we will have more to discuss on tables later). A particularly interesting RNS structure was introduced by Taheri [22] in 1988. This allows linear systolic arrays to be implemented with unary operation look-up tables, and is particularly effective for fixed coefficient filters. It is shown that a complete inner product pipeline can be tested using only $m + \log_2 m + 3$ vectors, where m is the ring modulus [23]. Note that this figure is independent of the number of stages, it is only dependent on the size of the modulus.

The RNS is only one of the methods available to provide direct product mappings. Another approach, is the polynomial residue number system (PRNS) [24], whereby a quotient ring is formed as the quotient of a formal-polynomial ring by an ideal generated by a polynomial which splits completely over either the base ring or an extension of it. For example, under certain circumstances, the ring $(Z_M[X])/(F(X))$ is isomorphic to the direct product of many copies of Z_M, the number of copies being equal to the degree of the polynomial F. Other methods utilize a type of overspanned complex number representation to develop direct product rings based on cyclotomic polynomial reduction [25][26]. A recent method uses direct mappings from the bit patterns of the data stream to generate finite ring polynomials that are mapped to a direct product ring [27]; an extension of this technique embeds very small rings of an RNS representation to provide large dynamic range computations using direct products of only 3-bit rings [28].

A final point to mention in the mapping to direct product rings, is the natural fault tolerance that can be achieved using redundant rings. Since the computations within one ring are independent of any other ring (no weighting information across the rings) then we can arbitrarily choose a sub-set of the rings with which to reconstruct the result. Assuming that the sub-set provides sufficient dynamic range for the intended calculation, we have several sub-sets from which to choose, and any sub-set will provide the correct result, assuming no errors in the computations. By comparing across the rings, we may isolate faulty rings and then remove them from the computational stream [29]. Redundant rings are used for both detecting and correcting errors, but it is also possible to separate the detection into a purely circuit technique, and allow the correction to use a smaller set of rings than would otherwise be possible [22].

Complex Arithmetic

The issue of complex arithmetic is normally not stressed in texts on computer arithmetic, since it is assumed that we will simply invoke the equations relating complex operations to real operations. The problem then returns to one of constructing ordinary adders and multipliers. We raise the issue here because many signal processing systems inherently deal with data streams that are best represented as complex numbers, and there are techniques available to reduce both the arithmetic and structural complexity of the processors. A typical example is in radar signal processing where signal delays and Doppler frequency shifts are naturally represented by complex signals [30], there are also many examples in communication systems where phase-shifts carry signal information [31] and in beamforming for antenna arrays [34]. The DFT, of course, is defined over the complex field.

In performing arithmetic on complex signals, we have a natural separation of real and imaginary components. Over addition, these components remain separated, but over multiplication, the components are mixed. If separate processing channels are desired for both addition and multiplication, then it is possible to define finite rings into which we may map the original real and imaginary channels and, providing we stay in these rings, there will be complete separation between the mapped components. We also have the added advantage of removing some arithmetic complexity from the processing, and the previously stated advantages associated with finite ring computations.

The QRNS is based on the principles of the RNS with the application of a special ring structure to allow the emulation of complex arithmetic calculations [32]. The Quadratic Residue Ring is defined by using the solution of the polynomial $x^2 + 1 = 0$; $x = j = \sqrt{-1}$, such that $j \in \Re(m_k)$. j will have two solutions in $\Re(m_k)$, where it can be shown that m_k is a prime with the form given in eqn. (13).

$$m_k = 4 \cdot K_k + 1 \tag{13}$$

Effectively we represent j with an indeterminate x and then take the algebraic quotient modulo the polynomial $x^2 + 1$. We choose moduli $\{m_k\}$ such that $x^2 + 1$ factors over the ring Z_{m_k}. We now map the real and imaginary components of the signal stream as shown in eqn. (14).

$$\begin{aligned} A^o &= a^r + j \cdot a^i \\ A^* &= a^r - j \cdot a^i \end{aligned} \tag{14}$$

where a^r and a^i are the real and imaginary components of the mapped sample. A^o and A^* are referred to as the normal and conjugate mapped components [33]. We are now able to construct independent channels for the normal and conjugate components, with as much processing as allowed by the computational range before having to map back to the complex field. The computational overhead is reduced, since both addition and multiplication are defined as integer operations over the rings computing the normal and conjugate components.

We may also consider the use of a more flexible mapping procedure (the FMRNS [35]) that increases the number of channels to 3 but allows any odd modulus to be used. The method is similar to the QRNS, but with the difference that the polynomial used to generate the ideal in the quotient ring is not the "sensible" one, namely $x^2 + 1$, but rather one that is merely convenient. Let $g(X)$ denote the polynomial $g(X) = X(X^2 - 1)$. This polynomial has roots $0, \pm 1$. We define $FMR(m_k)$ by the mapping $A^o = \alpha^r, A^* = a^r + a^i$, and $A^+ = a^r - a^i$, with $A^o, A^*, A^+ \in \Re(m_k)$. This is evaluation of the polynomial $a^r + Xa^i$ at the *three* roots 0, $+1$ and -1 of $g(X)$. This evaluation map sets up an isomorphism between the quotient ring of polynomials modulo the ideal generated by $g(X)$, and the cross-product of the ring $\Re(m_k)$ with itself *three* times. Multiplications and additions are performed component-wise in these rings, just as in the other cases. The FMRNS imposes no requirements on the prime divisors of m_k other than that they be odd.

PRE-STORED COMPUTATIONS

Read-only-memories (ROMs) have been proposed as arithmetic elements since the early 70's. ROMs store unminimized truth tables, and are ideal for switching functions that have a small number of input bits and that do not readily decompose. The use of ROMs to implement adders is an unsuitable medium because adders have low gate count and are more efficiently implemented by direct logic mapping. ROMs are normally suggested when it is required to implement functions that are not obviously decomposable. For example, we may resort to a ROM to store a set of sine values for function evaluation purposes, or for implementing a particular type of rounding [1]. A marginal example (for which there was probably a 'hardware window' in the early 70's) is a small binary multiplier. Assume that the ROM stores a table that performs a 4x4 multiplication (a 256x8 ROM); we simply build an array of these small multipliers to construct an arbitrary size binary multiplier. The ROM requires 2^{11} storage transistor locations, and two 4:16 decoders for addressing a two-dimensional ROM array. An implementation using adders will require a 4x4 AND plane, 3 half adders 6 full adders and a fast 3-bit adder. Although on a strict transistor count, the adder implementation is ahead, on a silicon area measure it is not that clear. The transistor locations used for the ROM storage are very dense. For very low numbers of inputs, (say ≤ 5, the adder (or gate) implementation is definitely more efficient. For medium numbers of inputs (say up to 8-10) the ROM implementation is reasonably competitive. For larger numbers of inputs the N^2 complexity of the ROM dominates, and it is not a suitable solution.

Stored Logic

The utility of using ROMs to store 'logic' thus has a viability window approximately between 5 and 10 bits. There will be special circumstances, of course, for general tables that have no obvious decomposition, where we may consider larger address words, but not for parallel computation on chip. In terms of commercial use of stored tables as logic, the Xilinx™ family of Field-Programmable-Gate-Arrays [36] uses tables downloaded to 4-5 bit input RAMs. In this case a small RAM is more efficient than general purpose programmable gates, and any small truth table can be implemented without the need for minimization. The main design objective is to decompose a large logic function into an interconnection of similarly sized logic blocks. There are also recent indications of the use of logic operations within the RAM circuitry itself [37]. The concept is to utilize the very large look-up bandwidth present at the row select sense amp. level in 2-dimensional DRAM structures. The result is a type of SIMD architecture that can be used to perform simple, but massively parallel, processing on the RAM data.

In this section we will show that opportunities exist for including viably sized fixed look-up tables, in particular for arithmetic processing architectures. Our particular focus will be on two techniques that allow efficiencies for fixed coefficient algorithms. The first is the *Distributed Arithmetic* technique and the second is the RNS system that we have already initially explored.

Distributed Arithmetic

Distributed arithmetic is a mechanism for replacing multipliers by distributing the operation of multiplication over bit-level memories and adders. The initial work was independently introduced by Croisier [38] and Peled and Liu [39] for digital filter design, a comprehensive survey can be found in reference [40] and a detailed VLSI design methodology in [41].

The target application is general inner product computation using fixed point arithmetic. The inner product computation can be decomposed, at the bit level, for one of the vectors, $\mathbf{x}$, as shown in eqn. (15).

$$\mathbf{x}^T\mathbf{y} = \begin{bmatrix} 2^{B-1} & 2^{B-2} & \cdots & 1 \end{bmatrix} \begin{bmatrix} x_{11} & x_{12} & \cdots & x_{1B} \\ x_{21} & x_{22} & \cdots & x_{2B} \\ \cdots & \cdots & \cdots & \cdots \\ x_{n1} & x_{n2} & \cdots & x_{nB} \end{bmatrix} \begin{bmatrix} y_1 \\ y_2 \\ \cdots \\ y_n \end{bmatrix} \tag{15}$$

We may now block partition the binary matrix of $\mathbf{x}$ and the vector, $\mathbf{y}$, in both the word and bit dimensions [41], as shown in eqn. (16), where q is the number of bit partitions of $\mathbf{x}$ and r is the number of word partitions of $\mathbf{x}$.

$$\mathbf{x}^T\mathbf{y} = \begin{bmatrix} 2^{B-1} & 2^{B-2} & \cdots & 1 \end{bmatrix} \begin{bmatrix} \mathbf{X}^T_{11} & \mathbf{X}^T_{12} & \cdots & \mathbf{X}^T_{1r} \\ \mathbf{X}^T_{21} & \mathbf{X}^T_{22} & \cdots & \mathbf{X}^T_{2r} \\ \cdots & \cdots & \cdots & \cdots \\ \mathbf{X}^T_{q1} & \mathbf{X}^T_{q2} & \cdots & \mathbf{X}^T_{qr} \end{bmatrix} \begin{bmatrix} \mathbf{y}_1 \\ \mathbf{y}_2 \\ \cdots \\ \mathbf{y}_r \end{bmatrix} \tag{16}$$

We may now write the partitioned version of the inner product as shown in eqn. (17).

$$\mathbf{x}^T\mathbf{y} = \sum_{i=1}^{q} \sum_{j=1}^{r} 2^{(p(i-1))/q} \mathbf{u}_j(\mathbf{X}^T_{ij}) \tag{17}$$

The function, $\mathbf{u}_j(\mathbf{X}^T_{ij})$, is stored in a look-up table addressed by the bits of $\mathbf{X}^T_{ij}$. The inner product can now be implemented by shifting and adding the output of the tables. Based on this partition scheme, we may choose the various parameters so that the look-up table size of $2^{np/rq}$ words is within a reasonable range. We will select this range based on efficient address sizes for ROMs as discussed above.

Residue Arithmetic

We have already discussed various aspects related to RNS; we will now turn our attention to the role of look-up tables in implementing arithmetic operations using RNS mapping. We already know that RNS arithmetic provides independent calculations over direct product rings; we now have to choose the rings for suitable table sizes.

Recall that the computational dynamic range is given by $M = \prod_{i=1}^{L} m_i$, where the

$\{m_i\}$ are relatively prime moduli. If we wish to implement general binary (2-variable) operations, then the address range of the ROM is $2B$, where B is the width of the moduli in bits (assuming all moduli have the same B). For a maximum 10-bit address width, we are dealing with 5-bit moduli. If we list relatively prime moduli, starting at $2^5 = 32$, we have the set $\{32, 31, 29, 27, 25, 23, 19, 17\}$ which provides the equivalent of about 37-bits of dynamic range. A particularly efficient implementation is possible if we are willing to settle for about 19-bits of dynamic range with 10-bits of scaled output. We choose the first 4 moduli $\{32, 31, 29, 27\}$ and scale results (with a mixed radix output [6]) by 31x29 which leaves about 10-bits of dynamic range (32x27). The advantage of having the last modulus as a power of 2 is that the combination of the final 2 mixed radix digits provides a 10-bit binary output [43]. Details of scaling and decoding are available in many references on basic RNS theory (e.g. reference [6]). Assuming 8-bit data and 8-bit coefficients, we have a 3-bit growth 'cushion' in the dynamic range. Using signal energy growth as the criterion, as discussed earlier, we can probably perform inner products of between 15 and 20 terms within the dynamic range. If we allow a fixed coefficient vector, as with the distributed arithmetic approach, then a single 10-bit address ROM will act as a complete multiplier/accumulator for each ring. As a final point, it is possible to combine distributed arithmetic with RNS [42]; in this case the bit sequence partition is already established by the choice of moduli.

As an example of a complete DSP system using 1024x5 ROMs, a FIR filter, using the systolic array of Figure 5, is shown in Figure 6.

Figure 6. RNS Computational Elements Using 10-bit ROMs

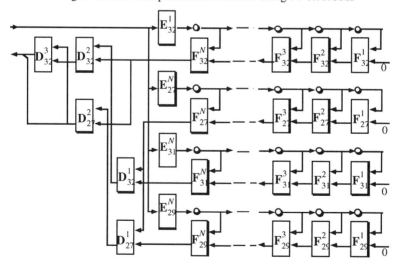

Each of the ROMs is assumed to contain pipeline latches on the 5 output bits. The systolic array structure can be clearly seen on the right hand side of the diagram (the circles are data pipeline latches), and the direct product ring exact replication of the basic architecture is clearly evident. Each ROM table is defined by the form $\mathbf{R}_{m_k}^{n}$, where $\mathbf{R}$ is the ROM function type (Filter, $\mathbf{F}$, Encoder, $\mathbf{E}$, and Decoder, $\mathbf{D}$), n is the sequence number of the ROM and m_k is the modulus of the ring calculation. The functions used for the various ROMs are given in eqn. (18), where x is the variable applied to the upper 5-bits and y is the variable applied to the lower 5-bits. The encoder is the exception, where the complete 10-bit input word is applied to the 10 address bits.

$$\mathbf{E}_{m_k}^{1} = (x)\,\mathrm{mod}\,m_k$$

$$\mathbf{F}_{m_k}^{n} = y\oplus_{m_k}\left(x\otimes_{m_k} W^n\right)$$

$$\mathbf{D}_{m_k}^{1} = \left\lfloor \frac{\left(\prod_{j=1}^{2} m_j\right)\left(x\otimes_{m_3}(\hat{m}_3)^{-1}\right)}{m_3} + \frac{1}{2}\right\rfloor \oplus_{m_k}\left(\frac{m_1\otimes_{m_k} m_2\otimes_{m_k}\left(y\otimes_{m_3}(\hat{m}_3)^{-1}\right)}{m_k}\right) \quad (18)$$

$$\mathbf{D}_{m_k}^{2} = \left\lfloor \frac{\left(\prod_{j=1}^{2} m_j\right)\left(x\otimes_{m_4}(\hat{m}_4)^{-1}\right)}{m_4} + \frac{1}{2}\right\rfloor \oplus_{m_k}\left(\frac{m_1\otimes_{m_k} m_2\otimes_{m_k}\left(y\otimes_{m_4}(\hat{m}_4)^{-1}\right)}{m_k}\right)$$

$$\mathbf{D}_{32}^{3} = (y\otimes_{32}(27)^{-1})\oplus_{32}(-x\otimes_{32}(27)^{-1})$$

The decoder is based on a scaled metric vector technique, that requires fewer ROMs, but with a slightly larger error than the more conventional exact division mixed radix technique. The scaled metric vector technique is well-established but does not appear to have been used widely [13][44].

It is also possible to decompose the inner product ROM, $\mathbf{F}$, into 5 cells, where each cell only requires a 5-bit addressed ROM. This technique has been well explored elsewhere and will not be discussed here [22][43][45][46].

Fast ROM Storage Mechanisms

A circuit element that we have relied on in many of the previous architectures, is the ROM cell. In our case the ROM is quite small (5K-bits) compared to current day storage requirements, and it is also assumed that the ROM is a pipelined element. Because of the small size we do not have to consider the problems with long bit-lines and with large decoders.

An initial step to constructing a small ROM is shown in schematic form in Figure 7. The circuit is for a single storage bit for a 5-bit addressed ROM. The circuit is dynamic and the pipeline is a 2-phase transmission gate dynamic master/slave latch. An interesting geometric layout is to group the transistors into a common row select structure (four transistors in each group) and to construct a STAR topology so that the drains may be merged [47]. Since the drains contribute most of the evaluation node parasitic capacitance, this represents an efficient layout strategy as far as pull down delay is concerned. A typical transistor grouping is shown in Figure 7 and the STAR layout style is shown in Figure 8. This grouping is not particularly efficient in area, but does allow for fast pulldown of the evaluation node because of the reduced load capacitance.

A complete 32 bit ROM array is shown in Figure 9; one of the 8 STAR structures is shown highlighted. The ROM is programmed by the selective inclusion of the transistor device well; the ROM configuration of Figure 9 is in an un-programmed state, so there is no transistor device well shown. We can see that the storage area is not as dense as we would expect for a normal ROM structure, but this ROM has been designed for maximum speed. The column select transistors are the wide geometries at the periphery of the storage array.

Figure 7. Pipelined Storage Bit for a 5-bit Addressed ROM

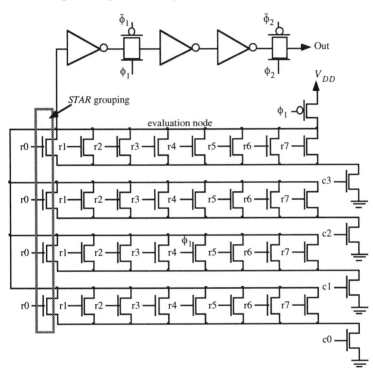

Figure 8. Star-Configured ROM

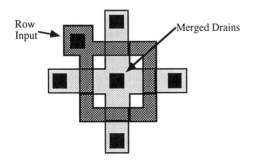

Figure 9. Complete 32-bit ROM Array

Switching Tree ROM Storage

We may trade the large capacitance of the dynamic bit line, for increased height of series transistors. The ROM configuration of Figure 7 is inherently 2-dimensional; that is, the address bits are split into two groups, each group is decoded and used to drive row and column select series transistors connected to the output bit-line. If we take the 5-bit address ROM as an example, there are a variety of dimensional decompositions we can consider. Table 1 shows the different ways

Factors	Dimension	Decode	Store	Total
32	1	165	160	**325**
2 16	2	69	170	**239**
4 8	2	37	180	**217**
2 2 8	3	29	190	**219**
2 4 4	3	21	210	**231**
2 2 2 4	4	13	230	**243**
2 2 2 2 2	5	5	310	**315**

Table 1. Decomposition Strategies for a 5-bit Address ROM

of factoring 32 to yield a variety of ROM layout dimensionality; an approximate idea of the complexity of each factor selection is given by total transistor count. We see that the (4,8) factor selection has the lowest number of transistors, and so seems to be the best. It is of interest, however, to examine other configurations for potential advantages that are not evident by initial transistor count. In order to examine the different configurations, we require a graphing technique to represent the different dimensionality of each set of factors. An elegant method is to use individual trees for each output bit.

The tree representations of the (4,8) and the (2,2,2,2,2) factor configurations are shown in Figure 10. We see that the dimensionality of the ROM is mapped to the tree height, and the decoder output for each dimension is mapped to the number of branches from the root nodes for each dimension. The height of the tree (or dimensionality of the ROM) is equal to the number of decoders that drive the ROM storage. As we increase the tree height, the decoders increase but become simpler. The (2,2,2,2,2) tree, a binary tree or decision tree, has decoders that are simple inverters. The ROM tree is programmed by removing branches from the lowest level of the tree, representing a 0 or 1 programmed at the particular location. The mapping of a logic 1 to the removal, or not, of a branch will depend on the final circuit realization. The full ROM will consist of B copies of the tree, individually programmed for each of the B output bits.

Figure 10. Tree Diagrams for the (4,8) and (2,2,2,2,2) ROM

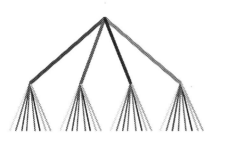 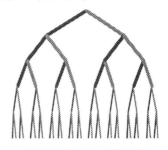

(4,8) Factor ROM Tree   (2,2,2,2,2) Factor ROM Tree

We can consider direct circuit implementation of any tree by mapping branches of the tree (or edges of the graph) to transistors, and driving the transistors with outputs from the decoders. In the case of the binary tree, we drive the "/" direction branches with the complement of the appropriate address bit and the "\" direction branches with the true bit. The advantage of using the binary tree is that, after programming has taken place, we may apply graph theoretical minimization rules prior to implementing the tree in circuitry.

An example of the minimization possible with this technique is shown in Figure 11, where the programmed function is $(a \times b) \bmod 7$.

Figure 11. Example minimization of a 64x3 ROM

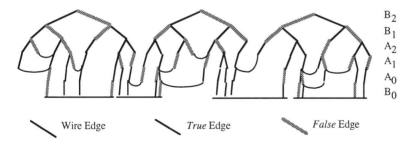

B_2
B_1
A_2
A_1
A_0
B_0

\ Wire Edge   \ *True* Edge   \ *False* Edge

The minimization procedure has been discussed elsewhere [48]; a VLSI module generator for automatically producing a tree layout has also been introduced in [49], and provides a direct mapping from the tree to silicon. The core layout of the Mod 7 module is shown in Figure 12. It is clear that the binary tree minimization has given us the opportunity for a large reduction in storage size (even though there is no obvious decomposition properties for a Mod 7 multiplier).

Figure 12. Core Layout of Mod 7 Multiplier Trees

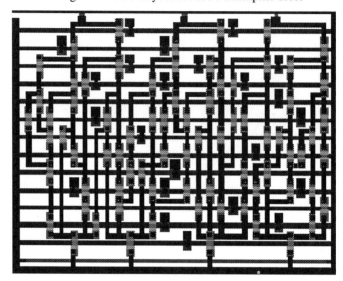

The core layout contains both the storage transistors and the distributed decoder structure (the 5 upper ROM dimensions). A direct transistor count is probably not fair, since the original storage ROM has fixed transistor positions, whether the position is occupied by a transistor, or not. A suitable 'counting' strategy here is to look at the effective transistor positions in the minimized trees. We basically count the number of tree branches that lead to a new column position in the array. This can be more clearly seen in the tree of Figure 11. If we use the algorithm that generates the tree, then the number of new columns is easily determined. The algorithm is given below:

1. Start at the top of the right hand tree; and map to the right most column in the matrix.

2. Move towards the bottom of the tree, taking either right hand edges or single merged edges, mapping the edges (vertical wire links or transistors) to matrix primitives in the column. Place horizontal wire matrix primitives if a previously mapped edge (in the right hand adjacent matrix column) is connected to the currently mapped edge. The path will terminate when either a left hand link is reached, or when the bottom of the tree is reached.

3. Move to the left until the first unplaced left hand edge, at any vertical position, is reached. Terminate the algorithm if all edges have been placed.

4. Repeat from 2, mapping to a new column to the left of the previous column.

Following this algorithm we find that 27 columns are required to generate the tree. We compare this to 3x64=192 transistor locations for the original ROM. The number of decoder and column select transistors (the top 5 levels of the tree) is 62. We can compare this to the 45 transistors required for the efficient (4,8) structure, and we see that the minimization has also reduced the tree very close to an equivalent decoder and column selection complexity.

FAST CMOS CIRCUIT TECHNIQUES

We have so far examined algorithmic and architecture issues, with some initial thoughts on circuit style. In this section we briefly review some factors that impinge upon fast arithmetic circuits. This is a large area of research and design, and we offer only a brief and very selective view in this section.

Circuit Style

Our focus is on CMOS and BiCMOS technologies as the vehicle for high performance arithmetic circuit design. The issue of the level of design abstraction for arithmetic circuits is important. Although VLSI design is moving to higher and higher levels of abstraction, the design of high performance circuitry is still a transistor level activity. Even gate level design styles are not going to provide the final edge to optimum performance.

Traditional CMOS design has concentrated on static logic gates with complementary blocks for pFET pull-up and nFET pull down functions [50]. Dynamic logic circuits have been demonstrated to provide faster evaluation and lower implementation area compared to their static counterparts [51], but industry was somewhat loathe to adopt the circuit style wholeheartedly. The situation has probably changed over the past few years as the requirement for orders of magnitude improvement in circuit performance for specialized systems has grown. The remainder of this section will concentrate only on dynamic logic styles to implement high performance arithmetic.

Algorithmic/Circuit Issues

To extract the last ounce of performance out of a circuit function, it is essential to link the circuit style to the algorithm or function being implemented. Often there is an attempt to look at performance from the point of view of gate delays, and also to estimate critical paths through a gate-level circuit by summing these delays. This is fine for a pessimistic estimate of performance, but architectures based on such calculations will not necessarily be optimum. This is particularly true for dynamic logic, since it is not obvious how to even measure the delay of an individual gate, particularly for relatively complex gates that are to be used in a domino logic type of chain. For short chains that are pipelined, then the complete critical path (if we can find it), including the latches, should be simulated at the transistor level, and techniques, such as transistor sizing, used to optimize the circuitry. There have been many recent introductions of dynamic logic families, and it is often difficult to decide which is the best choice; this is a decision that can only be arrived at by an exhaustive search of available techniques.

We will present an example, in the following section, of constructing a very fast binary adder from a special form of domino logic; this example illustrates the link between algorithm and circuit style, with non-obvious results for the critical path.

A Fast Adder

The work in this section has been the result of innovative algorithm/circuit developments from Prof. Zhongde Wang of the VLSI Research Group at the University of Windsor.

The basic algorithm for the adder reported here is the Carry-Look-Ahead (CLA) scheme. This is perhaps the most popular high performance adder technique and appears to have aggressive properties compared to other fast adder techniques

[54]. The main idea behind the CLA technique is to generate all carries for a particular block of adders (typically four) in parallel. The CLA carry information is produced from two considerations related to the addition operation of inputs x_i and at bit i. A carry out, c_{i+1}, is generated if both inputs are 1; an input carry is propagated if one of the inputs is 1. We write this as the Boolean expression shown in eqn. (19).

$$c_{i+1} = x_i \cdot y_i + c_i \cdot (x_i + y_i) = G_i + c_i \cdot P_i \tag{19}$$

G is known as the carry *generate* term and P as the carry *propagate* term.

If we expand this recursion to four stages, the expressions for the four carries are shown in eqn. (20).

$$c_1 = G_0 + c_0 P_0$$
$$c_2 = G_1 + G_0 P_1 + c_0 P_0 P_1$$
$$c_3 = G_2 + G_1 P_2 + G_0 P_1 P_2 + c_0 P_0 P_1 P_2 \tag{20}$$
$$c_4 = G_3 + G_2 P_3 + G_1 P_2 P_3 + G_0 P_1 P_2 P_3 + c_0 P_0 P_1 P_2 P_3$$

We see that the maximum fan-in of the AND gate implementation of each carry term is equal to the position of the carry. This normally poses a limitation on the number of adjacent stages that can be connected into a CLA network. It is possible to extend the hierarchy of this scheme, normally to one higher level. The generate and propagate functions are usually referred to as *Group Generate* and *Group Propagate* for this second level of hierarchy.

For the fast technique discussed here, we use the operator o, associated with the carry generation, introduced by Ladner and Fisher [55]. Assuming that two binary summands $a_n a_{n-1} \ldots \ldots a_1$ and $b_n b_{n-1} \ldots \ldots b_1$, are fed to the adder, with carry-in, c_0, assumed to be 0, we define the generate term, g_i, and propagate term, p_i, and exclusive OR, x_i, for each bit position i, as $g_i = a_i b_i$, $p_i = a_i + b_i$, $x_i = a_i \oplus b_i$.

The carry, c_i, and sum, s_i, are given by

$$c_i = (g_i, p_i) o c_{i-1}$$
$$= (g_i, p_i) o (g_{i-1}, p_{i-1}) o \ldots o (g_1, p_1) o c_0 \tag{21}$$

$$s_i = x_i \oplus c_{i-1} \tag{22}$$

where the operation, o, is defined by

$$(g_i, p_i) o (g_j, p_j) = g_i + p_i g_j + p_i p_j \tag{23}$$

$$(g_i, p_i) o c_{i-1} = g_i + p_i c_{i-1} \tag{24}$$

Note that o is an associative but not commutative operator [56]. Also note that p_i in eqn. (21), eqn. (23) and eqn. (24) can be replaced by x_i; but x_i in eqn. (22) can not be replaced by p_i.

The group generate, $G_{i,k}$, and group propagate, $P_{i,k}$, terms are given by eqn. (25)

$$(G_{i,k}, P_{i,k}) = (g_{i+k}, p_{i+k})o(g_{i+k-1}, p_{i+k-1})o...o(g_{i+1}, p_{i+1}) \qquad (25)$$

where the first subscript, i, indicates that the group starts from bit position $i+1$, and the second subscript represents the length of the group. $(G_{i,k}, P_{i,k})$ possesses the following properties.

$$
\begin{aligned}
(G_{i,k1+k2}, P_{i,k1+k2}) &= (G_{i,k2+k1}, P_{i,k2+k1}) \\
&= (G_{i+k1,k2}, P_{i+k1,k2})o(G_{i,k1}, P_{i,k1}) \qquad (26) \\
&= (G_{i+k2,k1}, P_{i+k2,k1})o(G_{i,k2}, P_{i,k2})
\end{aligned}
$$

$$(G_{i,1}, P_{i,1}) = (g_{i-1}, p_{i-1}) \qquad (27)$$

$$c_{i+k} = (G_{i,k}, P_{i,k})oc_i \qquad (28)$$

In our implementation we will use domino logic, which has the unfortunate drawback that logical inversion is not allowed [51], thus requiring both true and complement logic networks to be built. Alternatively, we may consider differential cascode voltage switch logic (DCVSL) circuits [57] where true and complement signals are generated within a single complex tree. Zhongde Wang's solution requires fewer transistors than either of these methods.

In order to generate the complement carry chain, we define a *pseudo-complement generate* $\hat{g}_i = \overline{a}_i \overline{b}_i$, and *pseudo-complement propagate* $\hat{p}_i = \overline{a}_i + \overline{b}_i$. Note that $\hat{g}_i$ is not the complement of g_i, and $\hat{p}_i$ is not the complement of p_i. Instead, $\hat{g}_i = \overline{p}_i$ and $\hat{p}_i = \overline{g}_i$. However, the new representation allows us to derive the complements of carries from $\hat{g}_i$, and $\hat{p}_i$ using the operator o.

The complement carry chain can be derived from $\hat{g}_i$; and $\hat{p}_i$ as shown in eqn. (29)

$$\overline{c}_i = (\hat{g}_i, \hat{p}_i)o\overline{c}_{i-1} = (\hat{g}_i, \hat{p}_i)o(\hat{g}_{i-1}, \hat{p}_{i-1})o......o(\hat{g}_1, \hat{p}_1)o\overline{c}_0 \qquad (29)$$

this equation shows an identical parallelism between the generation of carries and their complements. In a similar way, we define the *pseudo-complement group generate* $\hat{G}_{i,k}$, and the *pseudo-complement group propagate* $\hat{P}_{i,k}$, as shown in eqn. (30)

$$(\hat{G}_{i,k}, \hat{P}_{i,k}) = (\hat{g}_{i+k}, \hat{p}_{i+k})o(\hat{g}_{i+k-1}, \hat{p}_{i+k-1})o......o(\hat{g}_{i+1}, \hat{p}_{i+1}) \qquad (30)$$

The complement of the carry at the $i+k$ th bit position can thus be generated by eqn. (31).

$$\bar{c}_{i+k} = (\hat{G}_{i,k}, \hat{P}_{i,k})o\bar{c}_i \tag{31}$$

It is interesting to note that p_i in eqn. (29) and eqn. (30) can also be replaced by x_i. Unlike the conventional CLA adder, this method only requires a carry chain for a subset of bit positions.

Multiple output domino logic circuits (MODL), introduced by Hwang and Fisher [58], provide considerable hardware savings over single output domino logic circuits. The MODL concept relies on the observation that often an output logic function is the base of another output logic function. These two logic functions can be built on one single domino logic tree which produces two outputs. For the lookahead adder, using pseudo-complements, we can enhance the MODL concept (EMODL) to a more general case. If an intermediate logic function is the common factor of two or more output logic functions, these output functions can share a base subtree which implements the common factor.

In the carry and sum representations, eqn. (21), eqn. (22), and eqn. (30), it is clear that c_i and $\bar{c}_i$ are common factors for all carries and sums whose bit position is greater than i. For example, c_i and $\bar{c}_i$ are factors of s_{i+1}, c_{i+1} and $\bar{c}_{i+1}$; c_{i+1} and $\bar{c}_{i+1}$ are factors of s_{i+2}, c_{i+2} and $\bar{c}_{i+2}$, and so on. Thus, depending on the allowable height of the tree, several sums can be built on one tree. Figure 13, shows an example with a tree height of 5; four consecutive sums are built on one tree, with a single carry-in.

Figure 13. Four consecutive sums built on a single EMODL tree

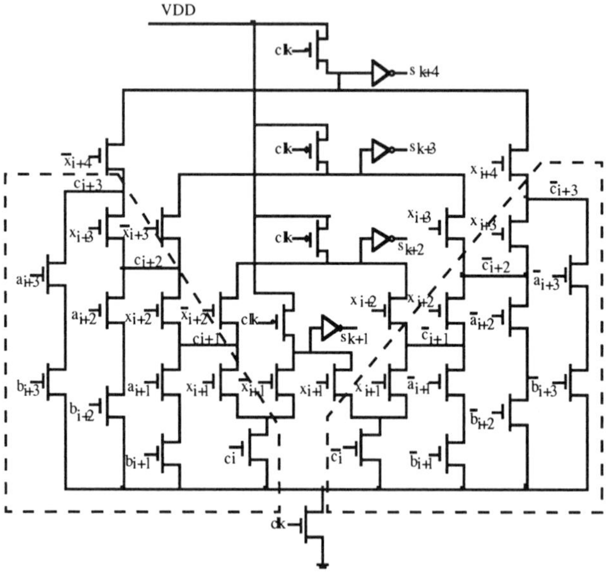

In Figure 13, the two trees, identified by the dashed blocks, are 4-bit carry chains. The left hand side block is the true carry chain; and the right side block is the complement carry chain. All four sums are built directly upon these two carry chain trees. The carries and their complements are indicated on the signal lines, but no output is produced for the carries. The tree height is limited to 5 based on considerations for charge sharing and pull down evaluation delay [51]. In this particular circuit, internal pull-ups are connected to c_{i+1} and $\bar{c}_{i+1}$ in order to reduce charge-sharing problems; the EMODL circuits are also sized in order to minimize pull down evaluation delay [59]. One of the advantages of the EMODL circuit approach is that only a few bottom transistors are required, and since these have the largest width when sizing the circuit, this tends to reduce the area required compared to other tree techniques.

We note that Figure 13 represents a full carry chain circuit. In the CLA style defined here, a full carry chain, which contains carries for all bit positions, can be replaced by a sparse carry chain, which contains only a small subset of the full carry chain. This reduces the load for building the carry chain, and thus contributes to a faster critical path. Since the full carry chain is not required by this approach, the bit level generates and propagates are not necessary. Therefore, the preliminary stage for any existing CLA adder, which produces bit level generates and propagates, can be eliminated. Instead of producing *bit level* generates and propagates, the first stage produces *group* generates and propagates. Eliminating the preliminary stage significantly contributes to critical path reduction.

A 32-bit adder has been reported from this work that has a critical path of 2.7ns [52] (verified by fabrication), and an analysis of a variety of architectures reveals that the number of cascaded stages is a more critical factor than the gate fan-in [53].

INTEGRATED DYNAMIC PIPELINES

The previous section has discussed fast circuit techniques, we now discuss high speed pipelined arithmetic. We will first make a case for dynamic logic pipelines and then discuss our particular approach to pipelined arithmetic, based on complex nFET trees. This somewhat of a departure from the very high speed pipelines that have been recently introduced, but suitable for most of the signal bandwidths that will be encountered in video-rate processing systems.

The Case for Dynamic Logic

There is a major advantage to using dynamic logic in any situation, and that is the complexity of dense transistor networks that can be accommodated within a single gate. The much lower input gate load and fast pull down, compared to static logic, is an additional benefit. In carrying the concept of dynamic logic to pipeline latches, we have the advantage of reduced transistor counts since we use charge storage on parasitic capacitances rather than the bistable nature of regenerative switches to hold data states. Disadvantages are the transient nature of parasitic storage, and the potential for charge change due to particle radiation effects. Traditionally, pipelines have been built using non-overlapping 2-phase clocks driving master/slave transparent latches. The basic concept is shown in Figure 7; the output bit latch for the pipelined ROM.

Recently published work has shown that it is possible to obtain extremely high pipeline rates (over 200MHz, for example, in a mature 3μ CMOS technology) by combining simple logic blocks with single phase clocked dynamic circuitry [60][61][62]. The trade-off in this approach is throughput rate versus latency (number of pipeline stages required). The ideal use for very high speed pipelines is in locally pipelined arithmetic units, where the local clock rate is much higher than the input data rate; bit-serial implementations are ideal target architectures. In the case of bit-level systolic arrays, the clock rate is the same as the rate of the

84

data stream, and we can therefore develop a circuit approach where there is a close match between the maximum throughput of the circuitry and the data stream rate. This is a particularly appropriate approach if a good hardware/speed trade-off is the result. The minimized tree ROM structures discussed earlier are ideal candidates for this moderated approach.

Pipelined Storage Tables

We embed a complex NFET logic block (switching tree) in a TSPC master/slave latch, as shown in Figure 14. Note that the p-channel logic block (highlighted) is restricted to a single PFET (inverter), operating as a slave latch. Our approach is to build the logic for each stage entirely within the NFET block; this provides the most area efficient implementation of a given logic function, and allows the use of an asymmetrical clock [63].

Figure 14. Embedded Tree in a TSPC Latch

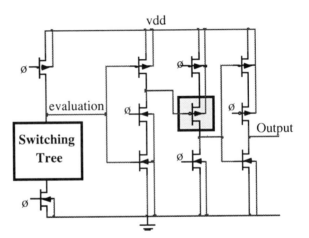

Trees built with this technique in a 3μ CMOS technology, have been clocked at 40MHz with a fan-in of 6. Our studies reveal that as technology density increases, the 'safe' fan-in of a minimized tree can be increased. Most of the problems associated with fan-in are associated with charge sharing [63], and this problem is mitigated as the ratio: $\dfrac{\text{drain capacitance}}{\text{incremental gate conductance}}$ decreases, as happens with decreasing device size. This effect points the way to increased tree fan-in and therefore more computational power per pipelined stage. It is important to point out that pipelining each stage reduces the charge-share problem significantly [63].

In the next section we consider the use of special tree sense circuits that can also contribute to increasing the tree height per pipelined stage.

Current Sensing

In this section we discuss a technique utilizing precharge/evaluate *current steering* to enable high fan-in blocks (of the order of 12) to be robustly pipelined in excess of 200MHz using a 1.2μ CMOS process.

Figure 15 Structure of Current Steering Latch

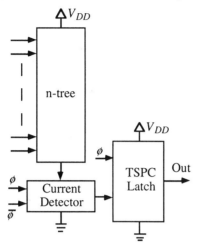

The approach is shown in block diagram form in Figure 15. One end of the tree is connected directly to power supply V_{DD}, the other end is directed to a current detector; the output of the tree is current. The input logic vector allows current through the tree, or not, based on the original truth table from which the tree is constructed. The output of the tree is detected by the current detector which, in the ideal case, has an internal resistance of zero. The output of the current detector is voltage which is Vdd for high and zero for low. The last stage in Figure 15 is the conventional TSPC latch. Note that the current detector requires a pseudo single-phase clock but this can generated under race-free conditions by a local inverter.

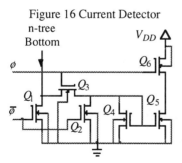

Figure 16 Current Detector

The circuit of the current detector is shown in Figure 16. The bottom of the tree is connected to the drain of Q_1 and Q_3; current conducted through the tree will be steered through Q_1 during pre-charge and through Q_3 during evaluate. During pre-charge, Q_2 will ensure that the current mirror, Q_4, Q_5, is off. During evalu-ate, any current conducted by the tree will be mirrored into the current/voltage converter, Q_6, Q_5, and then into the TSPC latch.

Note that we can reduce the charge-sharing and quadratic delay problems, associ-ated with switching high fan-in n-FET blocks, by designing the current steering circuit to limit the voltage swing at the bottom of the tree. We can also improve the latch input fall-time by ratioing the current mirror, Q_4, Q_5.

A typical pipelined output for a 12-high pipelined tree is shown in Figure 17. Note the low voltage swing at the base of the tree, limiting the voltage change across the tree. The elimination of the voltage swing required in the conventional latch input stage, realizes a doubling in pipeline frequency [64].

Figure 17. Output of a 12-high Pipelined Tree

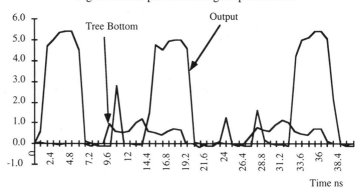

If we return to the initial discussion on storage elements for high speed arith-metic, we see that this pipelined tree structure is very suitable for both large dis-tributed arithmetic partitions and 5-6 bit finite ring computational elements for RNS implementations.

For interest, we show the layout of the tree portion (the majority of the layout area of a latched tree) of a 10-high tree implementing a Modulo 17 multiplica-tion; the layout can be seen in Figure 18.

Figure 18. 10-High Minimized Tree (Mod 17 Multiplier)

There are 1055 transistors, and 287 column branches. This is a considerable reduction from the 5120 transistor positions of a conventional ROM (plus decoder circuitry). The minimization procedure is clearly seen to have worked on the MSB of the 5-bit output (left hand side of the diagram). This bit has a value of 1 only for ring element of value 16. As we would expect, the tree structure to drive this bit is considerably smaller than the trees for the other bits. A typical tree minimization topology is seen in this example, i.e. the upper parts of the trees are quite sparse. It is quite possible that some of the gate drivers and latch circuitry can be built into this area of the trees. The overall dimensions of the tree are 1750μx170μ. Assuming about 1800 x 200 for a completely pipelined tree, we can fit over 200 cells on a single die for 1.2μ CMOS technology. If we take the RNS filter of Figure 6, this will allow an approximate filter size of over 45 taps with 20-bits computational range pipelined at a throughput rate (conservatively) of well over 100MHz. By any measure this represents fairly aggressive performance.

ARCHITECTURAL ISSUES.3 CASE STUDIES

As a final part to this chapter, we will present cameos of three case studies associated with high performance arithmetic, with DSP applications in mind. All three case studies are taken from previous or ongoing work at the VLSI Research Group, University of Windsor.

The first deals with regularity and interconnect length, which are important issues at the custom layout level of arithmetic design. The architecture concerned is a column compression multiplier (unsigned, or two's complement) which is optimally 'fast' but often not used because of the irregular nature of the layout and the interconnects.

The second is associated with the interface between RNS and binary number representations. We have already shown a special case that uses reasonably large look-up tables and 'tricks' limited to certain moduli sets; as a contrast, the technique discussed here uses small cells based on a binary arithmetic implementation of the Chinese Remainder Theorem.

The final study is taken from work on the construction of FFT butterfly elements for 2-D spectral analysis, using Quadratic Residue Rings. The architecture uses the ability of look-up tables to hide fixed coefficient multiplication. In this case, the butterfly is developed from an inner product realization of the butterfly computation, using more multiplications than the original! These, however, are hidden and the structure is enhanced.

Fast Multipliers

The fastest multipliers are obtained by arranging the compression of partial product bits in certain sequences. This type of multiplier is often referred to as a column compression (CC) multiplier. The principles for the CC multiplier were established by the early work of Ofman [65], Wallace [66], and Dadda [7]. It has been shown that the delay of the CC multiplier is proportional to $\log_{1.5} n$ [67], and the CC architecture is widely accepted as time optimal. The irregularity and complicated interconnections of the CC multiplier do not, however, readily allow efficient VLSI implementations, particularly for large n.

Dadda showed that different schemes, including the scheme proposed by Wallace [66], requires different numbers of cells (counters); the optimal scheme (requiring the least number of cells) is to compress the column size according to the recursion in eqn. ().

$$\sigma(0) = 2$$
$$\sigma(k+1) = \left\lfloor \frac{3}{2}\sigma(k) \right\rfloor \tag{32}$$

88

We can see in Figure 2 that column compression multipliers have irregular structures and complicated and long wiring connections.

In [68], Wang shows that it is possible to rearrange the ordering and connection of adders, to provide more regularity in the layout and more local interconnections. For short length multipliers it is also possible to reduce the length of the final fast adder. The new layouts are based on a heuristic procedure that takes into account bounds on the minimum number of adders and constraints associated with the allocation of adders to each stage in the multiplier, including an upper bound on adders in each stage. An example of an 8x8 multiplier design that regularizes the layout, provides nearest neighbour connections and reduces the length of the final fast adder, is shown in Figure 19.

Figure 19. Wang's 8x8 CC Multiplier

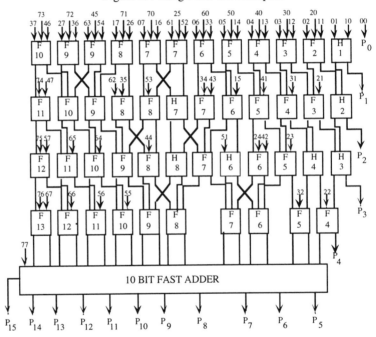

Further work shows that it is possible to extend the technique to two's complement multiplication [69].

Large Modulo Adders

In this work, we tackle the usually daunting problem of directly implementing the Chinese Remainder Theorem for a numbers of residues with a resulting large mapped ring modulus. Here, an algorithm for residue addition, based on a novel, 'non unique' number representation scheme, is implemented by a systolic array [70]. The array may be easily programmed by the user to accept any arbitrary modulus. Important applications of this array are in residue decoding and fault tolerant computation requiring the use of the Chinese Remainder Theorem where the modulus for addition is relatively large.

In conventional residue addition, a correction stage is involved whenever the sum exceeds the value of the modulus. This correction stage, a checking procedure, which determines whether or not the sum exceeds M, and subsequent addition of the correction, are all time consuming and hardware intensive. The situation is exacerbated for large values of M with awkward structures (many non-zero bits in a binary representation). Such adders can be considered expensive both in terms of critical path and silicon area.

The approach, presented by Bandyopadhyay [70], relies on constructing an adder where the size of the modulus is independent of the size of the primitive addition cell and number of pipeline stages required; also the throughput rate is maintained over the entire computational system and the adder structure is very regular. To meet these requirements, a systolic array is used for residue addition with carry save addition and a special non-unique RNS representation [71].

The basic pipelined multioperand adder is shown in Figure 20, using modulo M adders.

Figure 20. Pipelined Multioperand Mod M Adder

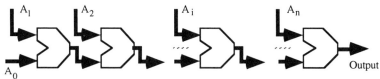

To represent the *ith* residue of X, the conventional method is to use p bits where $m_i \leq 2^p$. Since $0 \leq x < m$, the only valid bit configurations are those whose binary weighted values are less than m. Using this representation, a correction of $2^p - m$ is required; we usually select the correct result from a normal addition and a corrected addition.

The representation used by Bandyopadhyay assumes that all p-bit configurations, $2^{p-1} < m \leq 2^p$, are valid, and residues in the range $0 \leq x < 2^p - m$ will have two possible representations $\{x, x + m\}$. In effect we extend the redundancy of the carry-save representation to that of a redundant residue representation. The advantages of this non unique representation scheme are that corrections are applied only when there is a carry out of the MSB, and the operation can be pipelined [71]. We note that this non-unique representation has also been used in the work of Elleithy, et. al. [72].

We leave the details to the original references [70][71].

Semi-Systolic QRNS Butterfly

In this final cameo, we use the structural properties of the Quadratic Residue Number System (QRNS) to build a 2-D FFT butterfly using isolated normal and conjugate channels.

The computational element consists of a quadratic residue butterfly structure based on fixed coefficient inner product step processor (IPSP) cells (such as the minimized ROM cell discussed earlier). An IPSP cell is able to perform all the computational tasks demanded in the processor, including coding, butterfly computation, scaling and decoding.

There has been an exotic collection of algorithms and hardware structures produced over the past two decades on the subject of FFT implementations. Most of the algorithms are concerned with complexity reduction and mainly focus on reducing multiplication complexity [73]. The system discussed here is massively parallel, in that each twiddle factor multiplier can be regarded as a constant multiplier.

The use of the QRNS mapping allows independent calculations on the mapped normal and conjugate components. Consider the basic 2-D DFT operation on, say, the normal component, as shown in eqn. (33).

$$X^o_{k,l} = \sum_{n=0}^{1} \sum_{m=0}^{1} (x^o_{n,m} \otimes j^{2(nk+ml)} \otimes \alpha^o_{n,m}) \tag{33}$$

Here α^o is the normal component of the twiddle factor multiplier, with a premultiplication form for the butterfly. The above expression can be computed recursively (and a systolic structure obtained) using intermediate variables A and B as shown in eqn. ().

$$X^o_{k,l} = A^{[2k+1]} \oplus (\alpha^o_{1,0} \otimes j^{2k} \otimes B^{[2k+1]})$$

$$A^{[0]} = x^o_{0,0} + x^o_{0,1}$$

$$B^{[0]} = x^o_{1,0} + x^o_{1,1}$$

$$A^{[i+1]} = A^{[i]} \oplus \{x^o_{0,1} \otimes \alpha^o_{0,1} \otimes (j^{2(i+1)} \oplus [-j^{2i}])\} \tag{34}$$

$$B^{[i+1]} = B^{[i]} \oplus \{x^o_{1,1} \otimes \alpha^o_{1,1} \otimes (\alpha^o_{1,0})^{-1} \otimes (j^{2(i+1)} \oplus [-j^{2i}])\}$$

$$1 \le i \le 3$$

Note that we have defined an inverse, $(\alpha^o_{1,0})^{-1}$. Although we cannot rely on the QRNS mapped inverse for $\alpha^o_{1,0}$, the inverse exists for each component of $\alpha^o_{1,0}$, providing that we compute over a base field. A semi-systolic realization (with some broadcasting) of the butterfly is shown in Figure 21.

The butterfly consists of two arrays, each with four IPSP modules. Note that there is some interchange of data position and some broadcasting between the top and bottom two pairs of arrays. The basic pipeline structure, however, maintains two linear arrays running concurrently.

The function of the multipliers in each IPSP is both to provide the appropriate multiplier for the module and to cancel the unwanted term propagated from the previous module. The fact that multiplication does not involve extra hardware allows us to spread out the multipliers over the array. Contrast this to conventional structures that require fixed multiplication sites.

Note that there are actually more multiplications than in the original DFT butterfly, but this causes no real hardware overhead, providing the table sizes are the same for the different contents (remember that we minimize the tables based on pattern matching and this is a function of the contents of the table).

Figure 21. Semi-Systolic QRNS 2-D FFT Butterfly

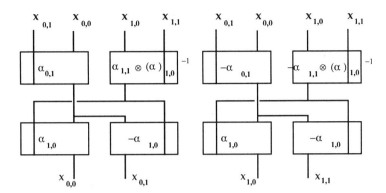

COMMENTS AND CONCLUSIONS

In this chapter we have discussed some issues relating to building high performance arithmetic elements, particularly directed at fixed point (integer) DSP applications. We have been interested in two different goals for the high performance circuits: fast (low critical path) and high speed (high throughput rate). We have taken a brief and somewhat tortuous path through the wealth of literature that is available on the subject. Our main aim has been to introduce different types of techniques that may be used to improve the performance of DSP arithmetic circuits and architectures. We have placed an emphasis on stored tables for computation with distributed arithmetic and RNS representations targeted for table look-up implementation. Our approach has been to address the search for high performance by looking at the areas of algorithms, architectures and circuit implementation of arithmetic elements. Finally we have presented three cameo studies showing different aspects of arithmetic processor design for DSP applications.

ACKNOWLEDGMENTS

The author wishes to acknowledge financial support from the Natural Sciences and Engineering Research Council of Canada, and the Micronet Network of Centres of Excellence. The loan of design hardware and software, and availability of fabrication services, from the Canadian Microelectronics Corporation was invaluable. Finally, the author wishes to acknowledge the support of the students and research staff of the VLSI Research Group, University of Windsor.

REFERENCES

[1]   Koren, I. *Computer Arithmetic Algorithms*. 1993, Prentice-Hall.

[2]   Scott, N.R. *Computer Number Systems and Arithmetic*. 1985, Prentice-Hall.

[3]   Swartzlander, E.E. ed. *Computer Arithmetic*. 1990, IEEE Computer Society Press Tutorial.

[4]   Blahut, R.E. *Fast Algorithms for Digital Signal Processing*. 1985, Addison-Wesley

[5]   Avizienis, A. "Signed-Digit Number Representations for Fast Parallel Arithmetic." *IRE Trans. Elec. Computers*, EC-10, 1961, pp. 389-400.

[6]   Soderstrand, M.A., Jenkins, W.K., Jullien, G.A., and Taylor, F.J., ed. *Residue Number System Arithmetic: Modern Applications in Digital Signal Processing*. 1986, IEEE Press.

[7]   Dadda, L. "Some schemes for parallel multipliers." *Acta Frequenza*, vol. 45, pp. 574-580, 1966.

[8]   Baugh, C.R., Wooley, B.A. "A Two's Complement Parallel Array Multiplication Algorithm". *IEEE Trans. Comput*. Vol. C-22, 1973, pp.1045-1047.

[9]   Booth, A.D. "A Signed Binary Multiplication Technique." *Q.J. Mech. Appl. Math*. Vol. 4, 1951, pp. 236-240.

[10]   Sam, H., Gupta, A. "A Generalized Multibit Recoding of Two's Complement Binary Numbers and its Proof with Applications in Multiplier Implementations." *IEEE Trans. Comput*., Vol. 39, No.8, 1990, pp. 1006-1015.

[11]   Swartzlander, E.E., Alexopoulos, A.G. "The Sign/Logarithm Number System", *IEEE Trans. Comput*. Vol. C-24, 1975, pp. 1238-1242.

[12]   Kingsbury, N.G., Rayner, P.J.W. "Digital Filtering uisng Logarithmic Arithmetic." *Electron. Lett*.,1971, No.7, pp.56-58.

[13]   Jullien, G. A. "Residue Number Scaling and Other Operations Using ROM Arrays." *IEEE Trans. Comput*. Vol. C-27, pp. 325-336, 1978.

[14]   Barraclough, S.R., Sotheran, M., Burgin, K., Wise, A.P., Vadher, A., Robbins, W.P., Forsythe, R.M., "The Design and Implementation of the IMS A110 Image and Signal Processor", *IEEE Custom Integrated Circuits Conf*., 1989, pp. 24.5.1-24.5.4.

[15]   Jenkins, W.K. "Composite number theoretic transforms for digital filtering." *Proc. 9th Asilomar Conference on Cir., Sys. and Comp*., Nov., 1975, pp. 458-462.

[16]   Jullien, G.A. "Implementation of Multiplication, Modulo a Prime Number, with Applications to Number Theoretic Transforms." *IEEE Trans on Computers*, Vol. C-29, No.10, October, 1980, pp. 899-905.

[17]   Baraniecka, A.Z., Jullien, G.A. "Residue Number System Implementations of Number Theoretic Transforms in Complex Residue Rings." *IEEE Trans. on Acoustics, Speech, and Signal Processing*, Vol. ASSP-28, No.3, June, 1980, pp.285-291.

[18]   Nagpal, H.K., Jullien, G.A., Miller, W.C. "Processor Architectures for Two-Dimensional Convolvers Using a Single Multiplexed Computational Element with Finite Field Arithmetic." *IEEE Trans on Computers*, Vol. C-32, No.11, November, 1983, pp. 989-1000.

[19]   Zelniker, G., F. J. Taylor. "A Reduced-Complexity Finite Field ALU." *IEEE Trans. on CAS.* Vol. 38, No. 12, 1991, pp.1571-1573.

[20]   Gold, B., Rader, C.M. *Digital Processing of Signals*, McGraw-Hill, New York, 1969.

[21]   Tretter, S.A. *Introduction to Discrete-Time Signal Processing.* John Wiley and Sons, New York, 1976.

[22]   Taheri, M., Jullien, G.A., Miller, W.C. (1988). "High Speed Signal Processing Using Systolic Arrays Over Finite Rings." *IEEE Transactions on Selected Areas in Communications, VLSI in Communications III*, Vol. 6, No. 3, April, pp. 504-512.

[23]   G.A.Jullien, M.Taheri, S.Bandyopadhyay, W.C. Miller, 1990, "A Low-Overhead Scheme for Testing a Bit Level Finite Ring Systolic Array." *Journal of VLSI Signal Processing*, Vol 2, No. 3, pp. 131-138.

[24]   Stouraitis, T., Skavantzos, A. "Parallel Decomposition of Complex Multipliers." Proc. 22d. Asilomar Conf. Circ. Sys. Comp., 1988, pp. 379-383.

[25]   Cozzens, J. H., L. A. Finkelstein. "Computing the Discrete Fourier Transform Using Residue Number Systems in a Ring of Algebraic Integers." *IEEE Trans. Inf. Th.* IT-31: 580-587, 1985.

[26]   Games, R. A. "An Algorithm for Complex Approximations in $Z[e2\pi i/8]$." *IEEE Trans. Inform. Th.* IT-32: 603-607, 1986.

[27]   N. Wigley, Jullien, G. A. "On Moduli Replication for Residue Arithmetic Computations of Complex Inner Products." *IEEE Trans. on Computers (Special Issue on Computer Arithmetic)* Vol. 39, No. 8, August, 1990, pp. 1065-1076.

[28]   Wigley, N.M., Jullien, G.A., Reaume, D. "Large Dynamic Range Computations over Small Finite Rings." *IEEE Trans. on Computers*, Vol. 43, No. 1, 1994, pp. 76-86.

[29]   Etzel, M.H., Jenkins, W.K. "Redundant Residue Number Systems for Error Detection and Correction in Digital Filters." *IEEE Trans. Acoust., Speech, Sig. Proc.* Vol. ASSP-28, 1980, pp.538-544.

[30]   McClellan, J.H., Purdy, R.J. "Applications of Digital Signal Processing to Radar." *Applications of Digital Signal Processing*, Oppenhein, A.V. ed., Prentice-Hall, New Jersey, 1978, Ch. 5.

[31]   Haykin, S. *Communication Systems.* John Wiley and Sons, New York, 1978.

[32]   Jenkins, W.K., Krogmeier, J.V. "The Design of Dual-Mode Complex Signal Processors Based on Quadratic Modular Number Codes," *IEEE. Trans. Circuits and Systems*, Vol. CAS-34, No. 4, April, 1987, pp. 354-364.

[33]   Jullien, G.A. Krishnan, R. Miller, W.C . "Complex Digital Signal Processing Over Finite Rings", *IEEE Trans. on Circuits and Systems*, Special Issue #5, Vol. CAS-34, No. 4, April, 1987, pp. 365-377.

94

[34]   Owsley, N.L. "Sonar Array Processing." *Array Signal Processing*, Haykin, S. ed., Prentice-Hall, New Jersey, 1985, Chapter 3.

[35]   Wigley, N.M. Jullien, G.A. "A Flexible Modulus Residue Number System for Complex Digital Signal Processing." *IEE Electronic Letters*, Vol. 27, No. 16, 1991, pp. 1436-1438.

[36]   *The Programmable Gate Array Data Book*, Xilinx Inc., 1992.

[37]   Elliott, D.G., Snelgrove, W.M. "CRAM: Memory with a Fast SIMD Processor." *Proceedings of the Camadian Conference on VLSI (CCVLSI'90)*, Ottawa, 1990, pp.3.3.1-3.3.6.

[38]   Croisier, A., Esteban, D.J., Levelion, M.E., Rizo, V. "Digital Filter for PCM Encoded Signals." *U.S. Patent 3777130*, December 3, 1973.

[39]   Peled, A., Liu, B. "A New Hardware Realization of Digital Filters." *IEEE Trans. Acoustics, Speech, Sig. Proc.*, Vol. ASSP-22, No. 6, 1974, pp. 456-462.

[40]   White, S.A. "Applications of Distributed Arithmetic to Digital Signal Processing: A Tutorial Review." *IEEE ASSP Magazine*, Vol. 6, No. 3, 1989, pp.4-19.

[41]   Burleson W.P., Scharf, L.L. "A VLSI Design Methodology for Distributed Arithmetic." *Journ. VLSI Sig. Proc.*, Vol. 2, No. 4, 1991, pp.235-252.

[42]   Jenkins, W.K. "Techniques for High-Precision Digital Filtering with Multiple Microprocessors." *Proc. 20th Mid-West Symp. Circuits. and Systems*, 1977, pp.58-62.

[43]   Jullien, G.A., "Number Theoretic Techniques in Digital Signal Processing", *Advances in Electronics and Electron Physics*, Academic Press Inc., vol. 80, Chapter 2, 1991, pp. 69-163.

[44]   Kameyama, M., Higuchi, T. "A New Scaling Algorithm in Symmetric Residue Number System Baesd on Multiple-Valued Logic." *Proc. Int. Symp. Circuits and Systems*, 1979, pp. 189-192.

[45]   Paliouras, V., Soudris, D., Stouraitis, T. "Systematic Derivation of the Processing Element of a Systolic Array Based on Residue Number System," *Proc. of Int. Symp. on Circ. and Syst.*, 1992, pp.815-818

[46]   Wolter, S., Klaasen, R., Birreck, D., Laur, R.. "VLSI Architecture for a Convolution-Based DCT in Residue Arithmetic," *Proc. of Int. Symp. on Circ. and Syst.*, 1992, pp.2108-2111.

[47]   Del Pup, L. "The Development and Application of High Speed Digital Switching Trees for Regular Arithmetic Arrays." *M.A.Sc. Thesis*, University of Windsor, Ontario, Canada, 1991.

[48]   G.A. Jullien, W.C. Miller, R. Grondin, Z. Wang, D. Zhang, L. Del Pup, S. Bizzan. "WoodChuck: A Low-Level Synthesizer for Dynamic Pipelined DSP Arithmetic Logic Blocks." *Proc. IEEE Int. Symp. on Circ. and Syst.*, Vol. 1, 1992, pp. 176-179.

[49]   Jullien, G.A., Wigley, N.M., Miller, W.C. "VLSI Implementations of Number Theoretic Techniques in Signal Processing", *Integration, the VLSI Journal*, (In Print)

[50]   Weste, N., Eshraghian, K. "Principles of CMOS VLSI Design." *Addison-Wesley*, Massachesetts, 1985.

[51]   Shoji, M. *CMOS Digital Circuit Technology*, Prentice-Hall, N.J. 1988.

[52]   Zhongde Wang, G.A. Jullien, W.C. Miller, Wang, J. "New Concepts for the Design of Carry Lookahead Adders", *Proc. IEEE Int. Symp. on Circ. and Syst.*, Chicago, May, 1993, pp. 1837-1840.

[53]   Wang, J., Zhongde Wang, Jullien, G.A., W.C. Miller. "Area-Time Analysis of Carry Lookahead Adders Using Enhanced Multiple Output Domino Logic", *Proc. IEEE Int. Symp. on Circ. and Syst.*, London, June, 1994 (in print).

[54]   Callaway, T.K., Swartzlander, E.E. "Estimating the Power Consumption of CMOS Adders." *Proc. of the 11th IEEE Int. Symp. on Comput. Arith.*, Windsor, Canada, 1993, pp.210-216

[55]   Ladner, R. E.,Fischer, M. J. . "Parallel prefix computation", *J. ACM*, vol. 27, pp.831-838, 1980

[56]   Chan, P. K., Schlag, M. D. F. "Analysis and design of CMOS Manchester adders with variable carry-skip", *IEEE Trans. Comput.* vol. C-39, pp. 983-992, 1990.

[57]   Chu, K. M., Pulfrey, D. I. "Design procedures for differential cascode voltage switch circuits", *IEEE J. Solid State Circuits*, vol. SC-21, pp. 1082-1087, 1986

[58]   Hwang, I. S., Fisher, A. L. "Ultra fast compact 32-bit CMOS adder in multi-output domino logic", *IEEE J. Solid State Cir.*, vol. 24, pp. 358-369, 1989.

[59]   Bizzan, S.S. Jullien, G.A. Miller, W.C. "Analytical Approach to Sizing nFET Chains", *IEE Electronics Letters*, vol. 28, No. 14, pp. 1334-1335, 1992.

[60]   Yuan, J., Svennson, C. "High-Speed CMOS Circuit Technique." *IEEE. J. Solid-State Circuits.* vol. 24 pp. 62-70, 1989.

[61]   Afghahi, M., Svensson, C. "A Unified Single-Phase Clocking Scheme for VLSI Systems." *IEEE J. Solid-State Circuits.* 25 Feb 225-233, 1990.

[62]   Yuan, J. Svensson, C.. "Pushing the Limits of Standard CMOS." *IEEE Spectrum*, February, pp. 52-53

[63]   Jullien, G. A., Miller, W.C., Grondin, R., Del Pup, L., Bizzan S., Zhang, D. "Dynamic Computational Blocks for Bit-Level Systolic Arrays," *IEEE Journal of Solid-State Circuits*, 1994, (in print).

[64]   Zhou, P., Czilli, J.C., Jullien, G.A.,Miller, W.C. "Current Input TSPC Latch for High Speed, Complex Switching Trees." *Proc. IEEE Int. Symp. on Circ. and Syst.*, London, June, 1994 (in print)

[65]   Ofman, Y. "On the complexity of discrete functions." *Soviet Physics-Doklady*, vol. 7, pp.589-591, 1963.

[66]   Wallace, C.S. "A suggestion for a fast multiplier." *IEEE Trans. Electronic Computers*, vol. EC-13, pp. 14-17, 1964.

[67] Cappello, P. R., Steiglitz, K. "A VLSI layout for a pipe-lined dadda multiplier." *ACM Trans. Comp. Syst.*, pp. 157-174, 1983

[68] Wang, Z., Jullien, G.A., Miller, W.C. "Column Compression Multipliers for Signal Processing Applications." *IEEE Workshop VLSI Sig. Proc.*, Nappa Valley, October, 1992.

[69] Wang, Zhongde, Jullien, G.A., Miller, W.C. "A New Design Technique for Column Compression Multipliers" *IEEE Trans. on Computers*, 1994, (In Print).

[70] Bandyopadhyay, S., Jullien, G.A., Bayoumi, M., 1986. "Systolic arrays over finite rings with applications to digital signal processing." *Proc. Systolic Array Workshop*, Oxford, pp. 123-131

[71] Bandyopadhyay, S., Jullien, G. A., Sengupta, A. "A Fast VLSI Systolic Array For Large Modulus Residue Addition." *Journ. VLSI Signal Processing*, 1994, In Print.

[72] Elleithy, K.M., Bayoumi, M., Lee, K.P. "O(logN) Architectures for RNS Arithmetic Decoding." *9th IEEE Symposium on Computer Arithmetic*, 1989, pp. 202-209

4

PIPELINING AND CLOCKING OF HIGH PERFORMANCE SYNCHRONOUS DIGITAL SYSTEMS

Eby G. Friedman* and J. H. Mulligan, Jr.**

*Department of Electrical Engineering **Department of Electrical and Computer Engineering
University of Rochester   University of California
Rochester, New York   Irvine, California
14627   92717

Abstract

This chapter discusses the effects of pipelining on the performance of high speed synchronous digital systems. In particular, the tradeoff between clock frequency and latency is described in terms of the circuit characteristics of a pipelined data path [1,2]. The design of the clock distribution network synchronizing the signal flow between each data path can significantly affect system performance. Timing characteristics of the clock distribution network are described in terms of how system performance can be either enhanced or degraded. A design paradigm relating latency and clock frequency as a function of the level of pipelining is developed for studying the performance of a synchronous system. This perspective permits the development of design equations for constrained and unconstrained design problems which describe these performance parameters in terms of the delays of the logic, interconnect, and registers, the nature of the clock distribution network, and the number of logic stages.

These results provide a new approach to the design of those synchronous digital systems in which latency and clock frequency are of primary importance. From the behavioral specifications for the proposed system, the designer can use these results to select the best logic architecture and the best available device technology to determine if the performance specifications can be satisfied, and if so, what design options are available for optimization of other system attributes, such as area and power dissipation. Approaches for exploring the effects of these other parameters on system performance are described. Furthermore, the results provide a systematic procedure for the design of a synchronous digital system once the logic architecture and technology have been selected by the designer.

1. Introduction

In the design of high performance synchronous digital systems, such as radar, sonar, and many types of high speed computers, there is a considerable desire for maximum performance. Performance, however, can be defined in many ways. Two key measures of performance are the latency of the system, which is the total time required to move a particular signal from the input of a system to its output, and the maximum clock frequency, which is measured by how often new data appear at the output of a synchronous digital system. In this chapter, performance is defined solely in terms of time and is represented by the latency and the clock frequency of the system. Thus, characteristics such as area or power dissipation are viewed as secondary design objectives.

Different applications of synchronous digital systems suggest different criteria for use in the optimization of their performance. For example, for a broad class of systems, optimization is done on the basis of a speed/area product. On the other hand, there are applications which are particularly sensitive to the latency of the system implementation. The results discussed in this chapter are primarily intended for feed-forward nonrecursive systems and describe a design approach for choosing the appropriate level of pipelining, thereby defining the system clock frequency and latency based on application-specific performance requirements and architectural and technological limitations. This chapter provides insight into speed/area/power tradeoffs within the system design space of a synchronous digital system. In addition, systems are discussed and examples are provided which support the use of these results.

In digital systems, the minimum latency occurs when the data path consists entirely of logic stages; it is the time required for propagation of a data signal through these logic stages. The clock period for this system, which is also the latency, is equal to the time required to process one data sample. If new data appear at the input of a system at time intervals smaller than the latency for this simple configuration, registers can be inserted into the data path to increase the frequency at which new data signals are processed through the system and appear at the system output. This degrades the latency, however. This process of inserting registers into a data path to increase the system clock frequency is spoken of as pipelining.

This chapter considers the design of systems in which one desires to maximize the clock frequency, minimize the latency, or achieve tradeoffs between minimum latency and maximum clock frequency. The chapter consists of six principal sections. In Section 2, relations between latency and clock frequency are developed in terms of the circuit and timing characteristics of a data path. The nature of the clock distribution network directly affects the performance and reliable operation of a synchronous digital system. Minimum and maximum constraint relationships, as well as design techniques to improve performance, are described in Section 3. A graphical interpretation of the performance tradeoffs of a pipelined feed-forward nonrecursive system is presented in Section 4, illustrating the constraints, limitations, and tradeoffs within the design paradigm of a synchronous digital system.

Most synchronous digital systems are designed to satisfy specific performance requirements such as minimum clock frequency or maximum latency. In these systems, the design problem is either one of maximizing the clock frequency while not exceeding a maximum latency or minimizing the latency while meeting a specified clock frequency. In certain systems, neither the latency nor the clock frequency ultimately constrains the design problem. In these unconstrained design problems, the level of pipelining of the data path can be chosen to trade off the latency with the clock frequency. These constrained and unconstrained systems are investigated and exemplified in Section 5. This design paradigm also supports the optimization of other performance parameters, such as area and power. In Section 6, approaches are described to integrate these additional design requirements into the system implementation. Finally, some conclusions are presented in Section 7.

2. High Performance Synchronous Data Paths

A synchronous digital system is typically composed of data paths in which data are moved from a register through some logic functions and into a second register. The synchronization of the data flow between the initial and final registers is typically coordinated by a single control signal, commonly called the clock signal. Thus, a synchronous digital system, as depicted in Figure 1, is composed of three interrelated systems:

1. the clock distribution network which generates the synchronizing clock pulse and defines when data can flow from one register to the next,

2. the registers which store the data signals awaiting the synchronizing clock pulse, and

3. the combinatorial network which contains the logical data paths of the digital system.

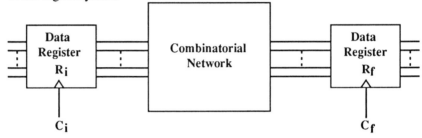

Figure 1: Synchronous Data Path

The total delay of a data path is determined by the time required to leave the initial register once a clock signal arrives, T_{c-Q}, the time necessary to propagate through the logic and interconnect, $T_{Logic} + T_{Int}$, and the time required to successfully propagate to and latch within the final register of the data path, T_{Set-up}. This relation is given below:

$$T_{PD} = T_{c-Q} + T_{Logic} + T_{Int} + T_{Set-up} \qquad . \qquad (1)$$

Equation (1) sums the individual delay components making up the total delay, T_{PD}, of any data path. The data paths whose total delay plus any clock skew are greatest represent the critical worst case timing requirements of a digital system, and the delay and clock skew of these paths must be minimized in order to maximize the performance of the entire digital system. Thus, in a high performance synchronous digital system, the critical paths constrain and define the maximum performance of the entire system. Therefore, the goal in designing a high performance system is to minimize each delay component in (1) as well as to utilize any possible advantages (and minimize any possible disadvantages) of the clock distribution circuitry which will increase the speed of operation of the critical data paths.

A general form of a data path is shown in Figure 2, where an initial register R_i begins the data path and is followed by N stages of logic and $N+1$ stages of interconnect, ending in a final register R_f. Each interconnect has been represented as a single pole time constant and designated as T_i, where i represents each logic and interconnect stage and N is the total number of logic stages. Thus, in a data path composed of N logic stages, there are $N+1$ interconnect time constants.

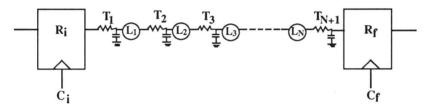

Figure 2: Synchronous Data Path with N Stages of Logic

Since logic paths are composed of only logic stages and interconnect sections, the total delay through a logic path can be modeled as the sum of the delay through the individual logic stages and interconnect sections. For convenience in representing the delay through the system, it is desirable to combine each individual RC interconnect section and logic stage as a single delay component of the logic path. This permits one to define the time required for the data signal to propagate through the ith distributed RC interconnect section T_i and logic stage L_i as T_{fi} and the average delay of all the logic and interconnect stages per data path as T_{fN}. Thus, an unpipelined data path provides the minimum latency L_{min}. For N logic stages traversed between the input and output of the system, L_{min} can be expressed as

$$L_{min} = \sum_{i=1}^{N} T_{fi} = NT_{fN} \qquad (2)$$

When registers are inserted into the data path from system input to output (defined as the global data path), the minimum clock period can be decreased (providing a higher maximum clock frequency), albeit with an increase in latency. Each global data path is composed of individual cascaded register-to-register data paths and these are defined as local data paths. Each local data path is composed of an initial and final register and typically, n logic stages between them. Note that each register within a local data path performs double duty, serving as the initial (final) and final (initial) register of the current and previous (next) local data path, respectively.

Each additional register generates delay components which are added to the logic and interconnect delays and which, when summed with the local clock skew, must be less than the clock period. Since there are fewer logic stages between registers in a pipelined data path than in an unpipelined data path, the clock rate is higher in a pipelined system. The register related delay components which are added to the path delay T_{PD}, as observed in Figure 2, originate in R_i and R_f. T_{c-Q} is the time interval between the arrival of the clock signal at R_i and the appearance of the data signal at the register output. The time required for the signal at the output of the final logic stage to propagate through the $n+1^{st}$ interconnect section and latch into the final register R_f is the set-up time T_{Set-up}.

The total delay from the output of the initial register R_i to the output of the nth logic stage is the sum of the individual T_{fi} terms along that data path as shown in (3) below:

$$T_{Logic} + T_{Int} = \sum_{i=1}^{n} T_{fi} \qquad . \qquad (3)$$

Thus, for a local data path consisting of n logic stages, the time delay through the path T_{PD} can be expressed as

$$T_{PD} = T_{c-Q} + \sum_{i=1}^{n} T_{fi} + T_{Set-up} \qquad . \qquad (4)$$

Equation (4) is composed of the delay required to get out of and into the initial and final registers, respectively, and the time required to propagate through n stages of logic and $n + 1$ sections of interconnect. If T_{Reg} represents the total register related delay of both R_i and R_f, then

$$T_{Reg} = T_{c-Q} + T_{Set-up} \qquad , \qquad (5)$$

and (4) can be written as

$$T_{PD} = T_{Reg} + \sum_{i=1}^{n} T_{fi} \quad . \qquad (6)$$

Thus, the total time to move a data signal through a data path is composed of the overhead requirements to get in and out of the register as well as the time required to perform the logical operations.

The maximum clock frequency at which a synchronous digital system can move data is defined in (7) as [1,2]

$$f_{clk} = \frac{1}{T_{CP}} \le \frac{1}{T_{PD} + T_{Skew}} \quad , \qquad (7)$$

where T_{CP} is the clock period, T_{PD} is defined in (6), and the local data path with the greatest $T_{PD} + T_{Skew}$ represents the critical path of the system, i. e., establishes the maximum clock frequency.

The latency L is defined as the time required to move a data signal from the input of the system to its output. For the special case of no pipelining, the latency equals the maximum clock period. If a single register is inserted into the data path, registers external to the system are required to synchronize the external data flow of the signal path. Two registers, one at the input and the other at the output of the global data path, represent a self-contained synchronous system (as shown in Figures 1 and 2). Each additional register increases the latency. Thus, the latency of a pipelined data path is the summation of the total delay through the global data path as shown below in (8) and (9).

$$L = \sum_{i=1}^{N} T_{fi} + \sum_{k=1}^{M} T_{ek} \quad , \qquad (8)$$

$$L = NT_{fN} + MT_{eM} \quad , \qquad (9)$$

where N is the number of logic stages per global data path, M is the number of local data paths (and clock distribution networks) per global data path, and $M+1$ is the number of clock periods (and registers) required to move a particular data signal from the input of the system to its output. The maximum permissible negative clock skew T_{ek} in (8) and (9) can be represented by (10). In that equation, T_{ek} is the aggregate delay of the kth local data path due to the initial and final registers (T_{Reg}) and the clock distribution network (T_{Skew}). T_e can be used to represent the margin of error or the acceptable tolerance of negative clock skew for each local data path. Note that when T_{Skew} is zero, T_{ek} equals T_{Reg}. Also note that T_{ek} is typically positive for most circuit configurations.

$$T_{ek} = T_{Reg} + T_{Skew} \tag{10}$$

T_e can be described as the total effective delay of the registers and clock distribution network per local data path. The average of the individual T_{ek} values over all the M serially connected cascaded data paths is defined as T_{eM}. Since T_{fN} is the average delay of each of the logic and interconnect stages (between the input and the output of the system), for convenience and improved interpretation (9) describes the latency in terms of average delays rather than individual summations.

The average number of logic stages per local data path n is given by

$$n = \frac{N}{M} \quad . \tag{11}$$

The clock period T_{CP} can be expressed as

$$T_{CP} \geq T_{Reg} + nT_{fn} + T_{Skew} \quad , \tag{12}$$

$$T_{CP} = \begin{cases} NT_{fN} & for\ M = 0 \tag{13} \\ T_{eM} + \dfrac{NT_{fN}}{M} & for\ M \geq 1 \quad . \tag{14} \end{cases}$$

For a pipelined global data path with registers placed at both its input and output, L can also be described by the relation,

$$L = \frac{M + 1}{f_{clk}} \quad . \tag{15}$$

Substituting (7), (10), and (12) into (15), the total latency of a partitioned global data path can be expressed in terms of the average delay components of a local data path as

$$L = (M + 1)(nT_{fN} + T_{eM})$$. (16)

These results assume that a global data path is partitioned into local data paths of approximately equal delay. By applying the concept of localized clock skew, which will be discussed in the following section, one can, in effect, even out the delays of each pipelined data path, making each local data path of approximately equal delay.

3. Clock Distribution Networks

In a synchronous digital system, the global clock signal is used to define a relative time reference for all movement of data within that system. Because this function is vital to the operation of a synchronous system, much attention has been given to the characteristics of these clock signals and the networks used in their distribution [1-22]. Most synchronous digital systems consist of cascaded banks of sequential registers with combinatorial logic between each set of registers. The functional requirements of the digital system are satisfied by the logic stages, while the global performance and local timing requirements are satisfied by the careful insertion of pipeline registers into equally spaced time windows to satisfy critical worst case timing constraints [1-5,10,22] and the proper design of the clock distribution network to satisfy critical timing requirements as well as to ensure that no race conditions can occur [3,5,6,22].

Each data signal typically is stored in a latched state within a bistable register awaiting the incoming clock signal to define when the data should leave the register. Once the enabling clock signal reaches the register, the data signal leaves the bistable register and propagates through the combinatorial network, and for a properly working system, enters the next register and is fully latched into that register before the next clock signal appears [4,8]. C_i and C_f represent the clock signals to the initial register and to the final register, respectively, and both originate from the same clock signal source. The times of arrival of the initial clock signal C_i and the final clock signal C_f shown in Figures 1 and 2 define the time reference when the data signals begin to leave their respective registers. These clock signals originate from a clock distribution network which is typically designed to generate a specific clock signal waveform which synchronizes each register [3-6,22]. The difference in delay between two sequentially-adjacent clock paths is described as the clock skew, T_{Skew}. If the clock signals C_i and C_f are in complete synchronism (i.e., the clock signals arrive at their respective registers at exactly the same time), the clock skew is zero.

The minimum allowable clock period between two registers in a sequential data path is given by

$$\frac{1}{f_{clkMAX}} = T_{CP_{min}} \geq T_{PD} + T_{Skew} \qquad , \qquad (17)$$

where T_{PD} is defined in (1) and (4) and T_{Skew} can be positive or negative depending on whether C_f leads or lags C_i, respectively. A timing diagram depicting each delay component in (1) in terms of the clock period is shown in Figure 3. These waveforms show the timing requirement of (17) being barely satisfied.

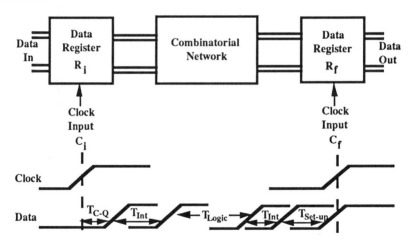

Figure 3: Timing Diagram of Clocked Data Path

3.1 Maximum Data Path/Clock Skew Constraint Relationship

For a design to meet its specified timing requirements, the greatest collective propagation delay of any data path between a pair of data registers, R_i and R_f, being synchronized by a clock distribution network must be less than the inverse of the maximum clock frequency of the circuit, as shown in (17). If the time of arrival of the clock signal at the final register of a data path, C_f, leads that of the time of arrival of the clock signal at the initial register of the same sequential data path, C_i, (see Figure 4A) the clock skew is referred to as a positive clock skew and, under this condition, the maximum attainable operating frequency is decreased [1-6,22]. Positive clock skew is the additional amount of time which must be added to the minimum clock period to reliably apply a new clock signal at the final register, where reliable operation implies that the system will function correctly at low as well as at high frequencies.

In the positive clock skew case, the clock signal reaches R_f before it reaches R_i. From (1), (7), and (17), the maximum permissible positive clock skew can be expressed as

$$T_{Skew} \leq T_{CP} - (T_{c-Q} + T_{Logic} + T_{Int} + T_{Set-up}) \qquad , \qquad (18)$$

where C_f leads C_i. This situation is the typical critical path timing analysis requirement commonly seen in most high performance synchronous digital systems. In circuits where positive clock skew is significant and (18) is not satisfied, the clock and data signals should be run in the same direction, thereby forcing C_f to lag C_i and making the clock skew negative.

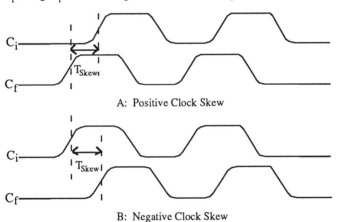

A: Positive Clock Skew

B: Negative Clock Skew

Figure 4: Clock Timing Diagrams

3.2 Minimum Data Path/Clock Skew Constraint Relationship

If C_f lags C_i (see Figure 4B), the clock skew is defined to be negative. Negative clock skew can be used to improve the maximum performance of a synchronous system by decreasing the delay of a critical path; however, a potential minimum constraint can occur [3,5,6,22], creating a race condition. In this case, the clock skew, when C_f lags C_i, must be less than the time required for the data to leave the initial register, propagate through the interconnect and combinatorial logic, and set-up in the final register (see Figure 3). If this condition is not met before the data stored in register R_f can be shifted out of R_f, it is overwritten by the data that had been stored in register R_i and had propagated through the combinatorial logic. Correct operation requires that R_f latches data which correspond to the data R_i latched during the previous clock period. This constraint on clock skew is given below:

$$|T_{Skew}| \leq T_{PD} = T_{c\text{-}Q} + T_{Logic} + T_{Int} + T_{Set\text{-}up} \qquad , \qquad (19)$$

where C_f lags C_i. In the negative clock skew case, the clock signal reaches R_i before it reaches R_f.

An important example in which this minimum constraint occurs is in those circuits which use cascaded registers, such as a serial shift register or a k-bit counter. As depicted in Figure 5, T_{Logic} is equal to zero and T_{Int} approaches zero (since cascaded registers are typically designed to abut at the geometric level). If C_f lags C_i (i.e., the local clock skew is negative), then the minimum constraint becomes

$$|T_{Skew}| \leq T_{c\text{-}Q} + T_{Set\text{-}up} \qquad . \qquad (20)$$

and all that is necessary for the system to malfunction is a poor relative placement of the flip flops or a highly resistive connection between C_i and C_f. In a circuit configuration such as a shift register or counter, where negative clock skew is a more serious problem than positive clock skew, provisions should be made to force C_f to lead C_i, as in the example shown in Figure 5.

As chips become functionally larger, on-chip testability is necessary. Data registers, configured in the form of serial set/scan chains when operating in the test mode, are a common example of a built-in test design technique [23]. The placement of these circuits is typically optimized around the normal operational data flow. When the system is reconfigured to use the registers in the role of the set/scan function, different delays are developed. In particular, the local clock skew can be negative and greater in magnitude than the local register delays. For example, with increased negative clock skew, (20) will not be satisfied and the incorrect data will latch into the final register of the poorly designed local data path.

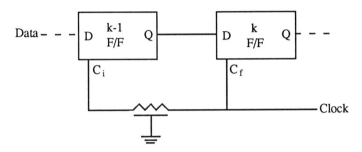

Figure 5: k-Bit Shift Register

Finally, in ideal scaling of MOS devices, all linear dimensions and voltages are multiplied by the factor 1/S, where S > 1. Device dependent delays, such as T_{c-Q}, T_{Set-up}, and T_{Logic}, scale as 1/S while interconnect dominated delays such as T_{Skew} remain constant to first order, and if fringing capacitance is considered, actually increase with decreasing dimensions [12,24,25]. Therefore, when considering scaling dimensions, (19) and (20) should be considered carefully.

3.3 Enhancing Synchronous Performance by Using Negative Clock Skew

Localized clock skew, as previously mentioned and as shown in (7) and (17), can be used to improve synchronous performance by minimizing the delay of the critical worst case data paths. By applying the approach of negative clock skew to the critical paths, excess time is shifted from the neighboring less critical local data paths to the critical local data paths by forcing C_i to lead C_f at each critical local data path. This negative clock skew represents the additional amount of time for the data signal at R_i to propagate through the n stages of logic and $n+1$ sections of interconnect and into the final register. Negative clock skew subtracts from the logic path delay, thereby decreasing the minimum clock period. This, in effect, increases the total time that a given critical data path has to accomplish its functional requirements by giving the data signal at R_i more time to propagate through the logic and interconnect stages and latch into R_f. Thus, the difference in delay of each local data path is minimized, thereby compensating for any inefficient partitioning of the global data path into local data paths, a common practice in most systems.

The maximum permissible negative clock skew of any data path, however, is dependent upon the clock period itself as well as the time delay of the previous data paths. This results from the structure of the serially cascaded global data path. Since a particular clock signal synchronizes a register which functions in a dual role, as the initial register of the next local data path and as the final register of the previous data path, the earlier C_i is for a given data path, the earlier that same clock signal, now C_f, is for the previous data path. Thus, the use of negative clock skew in the ith path results in a positive clock skew for the preceding path, which may establish the upper limit for the system clock frequency, as discussed below. For those consulting these references, it should be emphasized that in [3,5], the authors describe many of these characteristics of clock skew and its effects on the maximum clock frequency and designate the lead/lag clock skew polarity (positive/negative clock skew) notation as the opposite of that used here.

Example 1: The Application of Localized Negative Clock Skew to
Synchronous Circuits

Consider the non-recursive synchronous circuit shown in Figure 6, where the horizontal oval represents a logic delay and the vertical oval a clock delay. Since the local data path from R_2 to R_3 represents the worst case path (assuming the register delays are equal), by delaying C_3 with respect to C_2, negative clock skew is added to the R_2 - R_3 local data path. If C_1 is synchronized with C_3, then the R_1 - R_2 local data path receives some positive clock skew. Thus, assuming the register delays are both 2 ns., C_2 should be designed to lead C_3 by 1.5 ns., forcing both paths to have the same total local path delay, T_{PD} + T_{SKEW} = 7.5 ns. The delay of the critical path of the synchronous circuit is temporally refined to the precision of the clock distribution network and the entire system (for this simple example) could operate at a clock frequency of 133.3 MHz., rather than 111.1 MHz., if no localized clock skew is applied. The performance characteristics of the system, both with and without the application of localized clock skew, is summarized in Table 1.

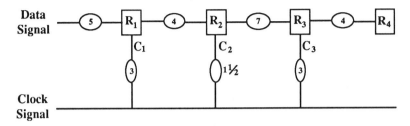

Figure 6: Example of Applying Localized Negative Clock Skew
to Synchronous Circuit

Note that $/T_{Skew}/ < T_{PD}$ (I-1.5 ns.I < 9 ns.) for the R_2-R_3 local data path; therefore, the correct data signal is successfully latched into R_3 and no minimum data path/clock skew constraint relationship exists. This design technique of applying localized clock skew is particularly affective in sequentially-adjacent temporally irregular local data paths; however, it is applicable to any type of synchronous sequential system, and for certain architectures, a significant improvement in performance is both possible and likely.

Table 1: Performance Characteristics of Circuit of Figure 6
Without and With Localized Clock Skew

Local Data Path	T_{PD}(min)-zero skew	T_{Ci}	T_{Cf}	T_{Skew}	T_{PD}(min)-non-zero skew
R_1 to R_2	$4 + 2 + 0 = 6$	3	1.5	1.5	$4 + 2 + 1.5 = 7.5$
R_2 to R_3	$7 + 2 + 0 = 9$	1.5	3	-1.5	$7 + 2 - 1.5 = 7.5$
f_{Max}	111.1 MHz.				133.3 MHz.

- all time units are in nanoseconds

The limiting condition for applying localized negative clock skew is determined by the control of the clock skew variations and by the difference in path delay between neighboring local data paths. These clock skew variations are due to process tolerances, where process parameters may vary over a specified range, and to environmental effects, such as temperature or radiation, which, for example, can both shift MOS threshold voltages and channel mobilities.

4. Design Paradigm for Pipelined Synchronous Systems

Registers are inserted into global data paths in order to increase the clock frequency of a digital system with, albeit, an increase in the latency. This tradeoff between clock frequency and latency is described in Figure 7 [1,2,22]. In this figure, both the latency and the clock period are shown as a function of the number of pipeline registers M inserted into a global data path. Thus, as M increases, the latency increases by T_{eM} for each inserted register, and the maximum possible clock frequency increases. This occurs because the critical path is shortened (since there are fewer logic and interconnect stages per local data path).

If no registers are inserted into the data path, the minimum latency L_{min} is the summation of the individual logic delays, $N\,T_{fN}$, as shown by (2) or by (13) with $M = 0$. As each register is inserted into the global data path, L increases by T_{eM}. Thus, L increases linearly with M; this is shown in Figure 7.

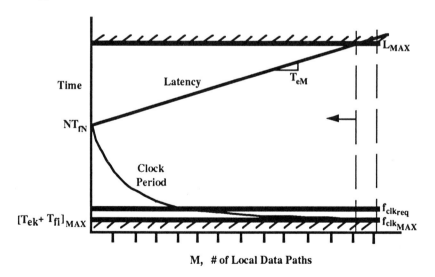

Figure 7: Design Paradigm for Pipelined Synchronous Systems

As seen from (14), the expression for the clock period contains a term which varies inversely with M; this behavior is shown in Figure 7. From (11) and (14), it is seen that the minimum clock period occurs when n equals one. The local data path having the largest value of $T_{ek} + T_{fi}$, defined as the critical data path, establishes the maximum clock frequency f_{clkMAX} for the system. This assumes that the circuit is not a simple shift register and logical operations are being performed between the registers of each local data path (if the global data path is a simple shift register, n equals zero and the maximum clock frequency is limited by the worst case T_{ek}). The MAX subscript in Figure 7 is used to emphasize that the critical local data path limits the minimum clock period (or maximum clock frequency) of the total global data path.

Most design requirements must satisfy some specified maximum time for latency while satisfying or surpassing a required clock frequency. The design constraints due to an application-specific limitation on the maximum permissible latency L_{max} and the maximum possible clock frequency f_{clkMAX} are shown in Figure 7 by the vertical dashed lines. Thus, for a given L_{max}, an appropriate maximum clock frequency and level of pipelining M is defined by the intersection of the L curve and the L_{max} line. If L_{max} is not specified or is very large and the desire is to make the clock frequency as high as possible, then an appropriate f_{clk} is defined by the intersection of the clock period curve and the f_{clkMAX} line. Thus, for a particular L and f_{clk}, the extent of the possible design space is indicated by the horizontal arrow. If L and f_{clk} are both of importance and no L_{max} or f_{clkMAX} is specified or constrains the design space, then some optimal level of pipelining is required to provide a "reasonably high" clock frequency while maintaining a "reasonable" latency. This design choice is represented by a particular value of M, defining an application-specific f_{clk} and L.

The effects of clock skew, technology, and logic architecture on clock period and latency are graphically demonstrated in Figures 8 and 9. If the clock skew is positive or if a poorer (i.e., slower) technology is used, then, as shown in Figure 7, T_{eM} increases and L reaches L_{max} at a smaller value of M than previously. Also, the minimum clock period increases which decreases the maximum clock frequency and which, for large positive clock skew or a very poor technology, eliminates any possibility of satisfying a specified clock frequency f_{clkreq} and limits the entire design space, as defined by the intersection of L and L_{max}. In addition, for a poorer technology or logic architecture, the intersection between either the clock period or the latency curve and the ordinate shifts upward since T_{fN} increases due to the slower technology and N increases for the less optimal architecture.

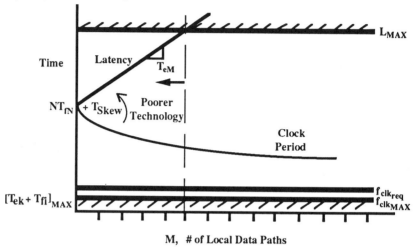

Figure 8: Effect of Positive Clock Skew and Technology
on Design Paradigm

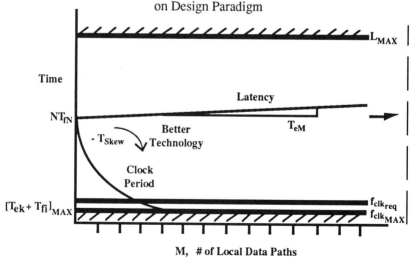

Figure 9: Effect of Negative Clock Skew and Technology
on Design Paradigm

If the clock skew is negative or a better (i.e., faster) technology is used, as shown in Figure 8, T_{eM} decreases and the latency is less dependent on M. Also, the minimum clock period decreases, satisfying f_{clkreq} and f_{clkMAX} with less pipelining. The possible design space, represented by the intersection of L and L_{max}, is greatly increased, permitting higher levels of pipelining if very high clock rates are desired. In addition, for a better technology or logic architecture, the intersection between either the clock period or the latency curve and the ordinate shifts downward since T_{fN} decreases due to the faster technology and N decreases for a more optimal architecture.

Thus, Figures 8 and 9 graphically describe how clock skew, technology, and logic architecture affect both the latency and the maximum clock frequency of a pipelined synchronous digital system. Finally, for applications which are not limited by the maximum latency or required clock frequency, it is shown in Section 6 how this design paradigm can be used to optimize performance parameters, such as area and power dissipation, while still satisfying the system specified clock frequency and latency requirements.

5. Design Objectives

Three types of design problems are considered using this approach [1,2,22]: 1) the maximum latency constrains the design problem, 2) the required clock frequency constrains the design problem, or 3) the problem is unconstrained and a tradeoff between L and f_{clk} must be made.

5.1 Maximum Latency

In applications where the maximum latency of a system is specified and L_{max} constrains the design space, the degree of pipelining can be determined from (21), where T_{eM} is taken as the estimate of an average $T_{Reg} + T_{Skew}$.

$$M \leq \frac{L_{max} - NT_{fN}}{T_{eM}} \qquad . \qquad (21)$$

A range of possible values of clock frequency is defined in (22), where a value of M is determined from (21). The lower bound on clock frequency is due to the constraint on maximum latency and the upper bound is required to ensure that the appropriate data for the time period are latched into the register.

$$\frac{M}{L_{max}} \leq f_{clk} \leq \frac{M}{MT_{eM} + NT_{fN}} \tag{22}$$

Thus, as shown in Figure 7, for a given maximum latency and knowledge of the average logic, register, and clock delay characteristics of a global data path, the degree of pipelining and range of clock frequency can be directly determined. An example is provided below describing this latency constrained design problem.

Example 2: Determining Clock Frequency for a Specified Latency

This example assumes that the delay characteristics of a pipelined data path are known, and the focus of the problem is to determine the range of frequency at which a system should be clocked while not exceeding a specified latency goal. This example can be explained in the context of Figure 7, where L_{max} cannot be exceeded while providing as high a clock frequency as possible. Thus, the appropriate level of pipelining to maximize f_{clk} while satisfying the constraint on L is determined by the intersection of the L and L_{max} curves, defining both M and f_{clk}.

Equations (21) and (22) can be used to determine the appropriate number of registers and the range of possible clock frequency, respectively, where it is noted that T_{eM} is the average $T_{Reg} + T_{Skew}$ of each local data path along the global pipelined data path. Thus, for a 100 stage data path where the average stage delay T_{fN} and average register and skew delay T_{eM} are both 2 ns., f_{clk} and an upper limit for M can be directly determined for a given target L as shown in Figure 10. If L must be less than 300 ns., then the maximum number of local data paths (M), corresponding to the upper limit given by (21), is 50 (requiring 51 pipeline registers). Thus, two logic stages per local data path, $n = 2$, are appropriate for this 100 stage system. In order not to exceed the target latency of 300 ns., the data must flow from register to register a maximum of every 6 ns. $(T_e + 2\ T_{fN})$, for a minimum clock frequency of 166.7 MHz. If, however, M

= 25 is chosen instead of 50, from (22), 83.3 MHz. $\leq f_{clk} \leq 100$ MHz. in order for this system to latch the correct data (since $T_{CP} = T_e + 4T_{fN} = 10$ ns.) and satisfy the maximum latency constraint.

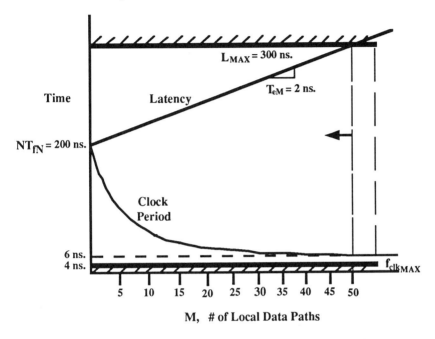

Figure 10: Example of Design Paradigm with Constraining Maximum Latency

5.2 Required Clock Frequency

In applications where the maximum clock frequency is specified and f_{clkMAX} constrains the design space, the latency and the number of registers can be determined from (9) and (23), respectively, where T_{CP} is the clock period defined in (12) - (14).

$$M = \frac{NT_{fN}}{T_{CP} - T_{eM}} \tag{23}$$

Thus, as shown in Figure 7, for a given maximum or required clock frequency and knowledge of the average logic, register, and clock delay characteristics of a global data path, the minimum latency and the required level

118

of pipelining can be directly determined. An example is provided below describing this clock frequency constrained design problem.

Example 3: Determining Latency for a Specified Clock Frequency

This example describes how the latency is determined for a specified clock frequency (or maximum clock frequency). This example can be explained in the context of Figure 7, where in this case the maximum clock frequency, not the maximum latency, constrains the design space. The appropriate level of pipelining to minimize L while satisfying f_{clkMAX} is determined by the intersection of the clock period and the f_{clkMAX} curves, defining both M and L.

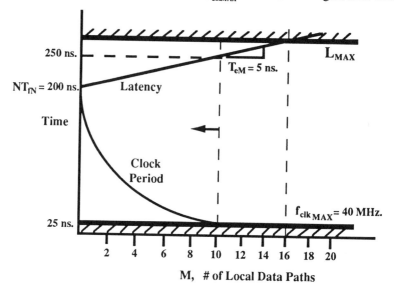

Figure 11: Example of Design Paradigm with Constraining
Clock Frequency

Equations (9) and (23) can be used to determine the specific latency and the number of registers, respectively, for a given maximum clock frequency. Assume a 100 stage data path with an average stage delay T_{fN} of 2 ns. and an average register and skew delay T_{eM} of 5 ns. is to be pipelined. L and M can be directly determined for a specified clock frequency, as shown in Figure 11.

If the maximum clock frequency is 40 MHz. (the clock period is 25 ns.), then the number of local data paths (M), given by (23), is 10 (requiring 11 pipeline registers). These ten local data paths add 50 ns. to the latency of the global data path, defining a total latency of 250 ns.

5.3 Unconstrained Design Requirement

Every register added to a data path increases the latency of a system by the added time required to move the data signal in and out of the register, T_{Reg}. This is typically accepted in order to increase the system clock frequency [25-32]. However, as additional stages of pipelining are inserted into a data path, the marginal utility of the increased clock rate is approached. This occurs when the time required to move data in and out of the register becomes comparable to or greater than the time incurred performing the logic functions.

Each additional register increases L by T_{eM} and decreases the maximum clock period by the decreased logic delay of the critical path. There exists an application-specific level of pipelining where the increase in latency costs the system more than the increase in clock frequency benefits the system. In order to quantify this, an arbitrary performance criterion (the pipelining efficiency, P_e) is defined to describe the performance cost of latency. P_e is a measure of the relative performance penalty incurred by the insertion of a single additional pipeline register to an existing global data path. It is a normalized function, shown in (24), which is the ratio of the total local logic delay to the total local data path delay, after the registers have been inserted. It defines what percentage of the local data path delay is logic related and what percentage is register related. As n increases, the ratio of the total local logic delay to the total local data path delay increases toward unity, reaching it when n is infinite (or practically, when the total local logic delay is much greater than the register delay).

$$P_e = \frac{\sum_{i=1}^{n} T_{fi}}{T_{PD}} = \frac{\sum_{i=1}^{n} T_{fi}}{\sum_{i=1}^{n} T_{fi} + T_{Reg}} \qquad . \qquad (24)$$

The benefit of inserting a register into a data path is increased clock frequency. If one starts with (7), a new relationship for clock frequency can be expressed as

$$f_{clk} \leq \frac{1}{T_{Reg} + \sum_{i=1}^{n} T_{fi} + T_{Skew}} = \frac{1}{\sum_{i=1}^{n} T_{fi} + T_e} \qquad . \qquad (25)$$

A measure of the cost/benefit of inserting registers into an N stage global data path is the function $P_e f_{clk}$, where P_e increases for increasing n, and f_{clk} decreases for increasing n. $P_e f_{clk}$ is thus a figure of merit for representing the performance advantages and disadvantages of pipelining. A different cost/benefit function could be applied if the effects of increased area, for example, were also of significant importance [29-31,34]. Cappello, LaPaugh, and Steiglitz [30,31] use an AP-product as a figure of merit (A is the chip area and P is the clock period) to optimize the speed/area performance of a pipelined system. The results described here, however, emphasize optimal latency and clock frequency over area/speed optimization. These results, coupled with the approaches described in Section 6 for analyzing the effects of other performance parameters on the design space, provide a systematic methodology for designing those systems in which both latency and clock frequency are of primary importance, and other key parameters, such as area and power dissipation, are of secondary importance.

If (24) is combined with (25), the figure of merit $P_e f_{clk}$ can be expressed as

$$P_e f_{clk} = \frac{nT_{fN}}{(nT_{fN} + T_{Reg})} \cdot \frac{1}{(T_{Reg} + nT_{fN} + T_{Skew})} \qquad . \qquad (26)$$

An optimal number of logic stages between pipeline registers N_{opt} is obtained from (26) and (27) by determining where the product $P_e f_{clk}$ is maximized or where

$$\frac{d(P_e f_{clk})}{dn} = 0 \qquad . \qquad (27)$$

By the use of (27), N_{opt}, the optimal number of logic stages per local data path for a high speed synchronous digital system, is obtained as

$$N_{opt} = \frac{1}{T_{fN}} \sqrt{T_{Reg} (T_{Reg} + T_{Skew})} \qquad . \qquad (28)$$

Under the condition of an ideal clock distribution network with zero clock skew, (28) simplifies to

$$N_{opt} = \frac{T_{Reg}}{T_{fN}} \qquad . \qquad (29)$$

N_{opt}, in (29), is the ratio of the register delay overhead to the average logic and interconnect stage delay of the global data path. If T_{Reg} is much less than T_{fN}, which occurs when T_{fN} represents the delay of a large multi-level function, the cost of inserting registers is small and N_{opt}, from (29), should be as small as feasible (since N_{opt} must be an integer, its smallest realizable value is one) or one should pipeline as often as the system permits. If T_{Reg} is much greater than T_{fN}, which often exists when operating at the level of individual logic stages, then the cost of inserting registers is high and N_{opt} is some large number determined from (29). Another interpretation of (29) is that the optimal number of logic stages between registers occurs when the total logic path delay NT_{fN} equals the total register delay T_{Reg}, thereby maximizing $P_e f_{clk}$.

T_{Skew} in (28) can be positive, negative, or zero with the constraint that if T_{Skew} is negative, its magnitude must be less than T_{Reg}. It is interesting to note that the effect of clock skew on N_{opt} is dependent upon its relative magnitude with respect to T_{Reg} and T_{fN}. Thus, if T_{Reg} is large with respect to T_{Skew}, the relation essentially reduces to (29). Also, positive clock skew adds directly to T_{Reg} and increases the cost of pipelining, thereby increasing the recommended number of logic stages between registers and quantifying how the clock distribution network affects the optimal design of a high speed data path. Note that negative clock skew decreases the cost of pipelining as well as increases the clock frequency of the critical paths.

The performance of a pipelined synchronous system can be maximized when the cost of the registers is minimal, thereby permitting the frequent insertion of pipeline registers and a higher overall system clock rate. As implied by (28) and shown in Figure 9, if T_{Skew} is negative, the overall performance cost of pipelining decreases.

The maximum permissible negative clock skew in (28) can be represented by T_e, where T_e is the margin of error or the acceptable tolerance as defined by (10). Equation (28) can then be rewritten as

$$N_{opt} = \frac{\sqrt{T_{Reg} T_e}}{T_{fN}} \qquad . \qquad (30)$$

For a given N_{opt}, T_e, and the average gate delay T_{fN}, the optimal frequency at which a particular data path (and system) should operate and the optimal level of pipelining can be computed from (31) and (32), respectively. In an actual application, N_{opt} should be rounded to an integer value, hereby denoted as N^*_{opt}. Rounding to the next highest (lowest) integer value improves (degrades) the density since less (more) registers are used but decreases (increases) the clock frequency since there are more (less) logic stages per local data path.

$$f_{clkopt} \leq \frac{1}{T_{Reg} + N^*_{opt} T_{fN} + T_{Skew}}$$

$$= \frac{1}{N^*_{opt} T_{fN} + T_e} \qquad (31)$$

Note that for positive clock skew, the maximum optimal clock frequency decreases, while for negative clock skew the maximum optimal clock frequency increases.

$$M_{opt} = \frac{N}{N_{opt}^*} \qquad (32)$$

N_{opt}^* is the optimal integer number of logic stages within a critical data path and therefore, if N/N_{opt}^* is not an integer, some of the local data paths should contain $N_{opt}^* - 1$ logic stages. An example of this unconstrained design problem is provided below.

Example 4: Determining an Optimal Tradeoff Between Clock Frequency
and Latency

This example describes a design problem in which neither the clock frequency nor the latency is constrained by the circuit specifications. This example can be explained in the context of Figure 6 where the target latency and clock frequency are less than L_{max} and f_{clkreq}, respectively, and the design space is unconstrained. Therefore, a tradeoff between L and f_{clk} must be made.

Equation (28) can be used to determine the appropriate number of logic stages between registers. Thus, assume a 30 stage data path is under consideration with an average stage delay T_{fN} of 4 ns., an average clock skew of a positive 4 ns., and a register delay of 6 ns. Equation (28) applies and N_{opt} is obtained as shown below:

$$N_{opt} = \frac{1}{T_{fN}} \sqrt{T_{Reg} (T_{Reg} + T_{Skew})}$$

$$= \frac{1}{4} \sqrt{6 (6+4)} = 1.94$$

Therefore, since N_{opt} is 1.94, there should be two logic stages per pipelined data path. From (31), the optimal maximum clock frequency is 55.6 MHz. Since $N_{opt} = 2$ and $N = 30$, M_{opt}, from (32), is 15 local data paths (or 16 pipeline registers). Since $T_{eM} (T_{Reg} + T_{Skew})$ is 10 ns., the total latency of this system, from (9), is 270 ns. Thus, an optimal implementation of this high performance global data path is 15 local data paths operating at a maximum of 55.6 MHz. with a latency of 270 ns.

6. Incorporation of Additional Performance Criteria

In this chapter three types of design problems have been described, where the goal is either to provide the highest clock frequency without exceeding a maximum latency, to satisfy a required clock frequency while minimizing the latency, or to provide a reasonable tradeoff between both the latency and the clock frequency by applying the concept of marginal utility to both of these parameters. All of these strategies permit the incorporation of additional performance parameters and these added degrees of freedom can be described in the context of the design paradigm described in Section 4. A powerful approach for dealing with these design problems is the use of the pipeline factor M to drive the design process to reduce area or power consumption or both. This design approach is represented in Figure 7 by a choice of M less than that constrained by L_{max} or f_{clkMAX}. With this methodology, the effects of other performance goals on the design space can be explored while still satisfying an application-specific L and f_{clk}.

Different options are possible to improve the speed/area/power implementation of a pipelined synchronous digital system. These include 1) pipelining less frequently, 2) changing the nature of the logic stages, 3) changing the system architecture, or 4) a combination of the aforementioned design options. For example, pipelining less often saves register area and power dissipation. Alternatively, one can change the nature of the logic stages by either using smaller logic stages which consume less circuit area and source and sink less current, thereby dissipating less power or by selecting a less aggressive high speed technology which could save area and power dissipation and possibly save cost through increased yield. An example of the latter is using a standard silicon-based technology such as CMOS instead of a higher speed and more expensive technology, such as GaAs or High Electron Mobility Transistors (HEMT). Another option is to implement the system with a more power and area efficient lower speed architecture which still satisfies system specifications. Finally, by combining each of these design options, a highly desirable system implementation could result.

6.1 Pipelining Less Frequently

Pipelining is applied to increase system clock frequency. For those applications which require lower levels of clock frequency (i.e., the problem is unconstrained or the maximum latency constrains the design problem), a choice of M can be made which is less than that defined by L_{max} or f_{clkMAX}. If fewer registers are used since M is smaller, the total system area and power dissipation is decreased. For example, for a 100 stage global data path composed of 20 local data paths, if M is decreased by 20%, the total area and power dissipation for the complete system would decrease by approximately 7.5% and 4.5%, respectively. Thus, as M is decreased, the design space becomes constrained by the application-specific L_{max} and f_{clkMAX}. Since less area and power dissipation are required in this system implementation, this new choice of M represents an improvement as compared to the unconstrained design problem.

As M is decreased, the latency and clock period change (see Figure 7). L decreases and T_{CP} increases (the clock frequency decreases) for decreasing M, as shown in (9) and (14). T_{CP} increases since n, the number of logic stages between registers, increases for decreasing M, as shown by (11) and (12). Thus, the level of pipelining M can be decreased until either L equals L_{max} or T_{CP} equals T_{CPreq}, whichever comes first, where T_{CPreq} is the maximum required clock period (minimum required clock frequency). This constrains the design space and defines the appropriate level of pipelining. Design equations describing these conditions are

$$M_1 = \frac{L_{max} - NT_{fN}}{T_{eM}} \quad , \quad (33)$$

$$M_2 = \frac{NT_{fN}}{T_{CP_{req}} - T_{eM}} \quad , \quad (34)$$

where the appropriate level of pipelining M is defined by

$$M = Max \ \{M_1, M_2\} \quad . \quad (35)$$

Thus, for unconstrained design problems, the level of pipelining can be decreased, saving circuit area and power dissipation, until either the application defined maximum latency or the required clock frequency is reached. Once either constraining design condition, (33) or (34), is satisfied, the limiting application-specific level of pipelining M is defined.

6.2 Changing the Nature of the Logic Stages

If the problem is not constrained by the maximum latency or the required clock frequency, another approach for improving the speed/area/power dissipation of a particular system implementation is, for a given logic architecture (i.e., N is constant), to change the nature of the logic stages used. In an unconstrained design problem, different strategies can be used to optimize the delay of the logic stages. Either a less aggressive technology can be used or the logic stages can be designed to be slower (since T_{fN} can be larger), permitting the circuit area, for a given technology, to be smaller. These smaller logic stages source less current and therefore dissipate less power. The design paradigm graphically expresses the significance of this approach. Since T_{fN} increases, from (9) and (14), the latency and clock period also both increase. In Figure 6, both the latency curve and the clock period curve shift upward for increasing T_{fN}. Thus, T_{fN} can be increased until either L equals L_{max} or T_{CP} equals T_{CPreq}, whichever comes first, thereby constraining the design space and defining a more optimal level of average delay per logic stage. Design equations describing these conditions are

$$T_{fN_1} = \frac{L_{max} - MT_{eM}}{N} \quad , \quad (36)$$

$$T_{fN_2} = \frac{M}{N}(T_{cP_{req}} - T_{eM}) \quad , \quad (37)$$

where the appropriate stage delay T_{fN} is defined by

$$T_{fN} = Min\{T_{fN_1}, T_{fN_2}\} \quad . \quad (38)$$

Thus, for unconstrained design problems, a slower technology can be used or the logic stages can be designed to minimize area and power dissipation instead of minimizing delay. In either case, once a target logic stage delay is known, many approaches exist for making design tradeoffs among speed, area, and power dissipation for a variety of different technologies. Some examples of these approaches, in which the logic delay is described in terms of specific technological, geometric, and electrical characteristics, are provided in [35-46].

In either approach, whether a different technology is used or the circuit design of the logic stages emphasizes area and power over speed (or a combination of the two), the new system implementation has advantages for the particular application. While still satisfying the application specified latency and clock frequency requirements, less area and power are required.

6.3 Changing the System Architecture

Other possible approaches besides decreasing M or increasing T_{jN} can be considered for a given unconstrained system. For example, new system architectures can be considered which are slower but consume less area and power. This equates to decreasing N, the number of logic stages per global data path, but increasing T_{jN}. A simple example of this approach is replacing a high speed carry look ahead adder with a serial adder, assuming the added performance of the carry look ahead adder is not required. In this instance, the serial adder would be much slower than the carry look ahead adder but consume considerably less area and power. In the context of the design paradigm depicted in Figure 6 and described in Section 4, N would decrease since more of the function is done serially, but T_{jN} would increase since the total delay from system input to system output is greater, thereby increasing the average logic stage delay. This approach can be used until the increase in T_{jN} is significantly greater than the decrease in N, shifting both curves in Figure 6 upward until either the L_{max} or f_{clkreq} constraint is reached.

Any subset of these aforementioned approaches can be combined during the design phase in order to reach a more desirable system implementation. Thus, the design paradigm, coupled with design principles discussed in this section, can be used with application-specific performance constraints to explore the effects on system requirements of other performance parameters, such as area and power dissipation.

7. Conclusions

There is an important class of practical system applications, such as radar, sonar, and high speed computing, which requires both high clock frequency and minimal latency and for which there is a need for developing a design methodology which satisfies their application-specific performance objectives. The results presented in this chapter describe a systematic strategy for designing high performance systems in which both clock frequency and latency are of primary interest. Additional performance attributes, such as power dissipation and area, are also discussed and design approaches are described for those applications where the system requirements are not limited by the available design space.

In feed-forward nonrecursive systems, global data paths are often partitioned into local pipelined data paths, thereby decreasing the delay of the critical paths and increasing the clock frequency, albeit with an increase in latency. The graphical design paradigm presented herein permits one to analyze how the performance of a synchronous system is affected by its degree of pipelining. This perspective permits the development of design equations for describing pipelined data paths in terms of the logic, interconnect, and register delays, localized clock skew, the performance efficiency of pipelining, and the total number of logic stages per local data path.

Clock distribution networks are described in terms of their data path timing requirements. Clock skew was shown to affect performance by both its magnitude and its lead/lag relationship with its sequentially-adjacent clock waveforms. Localized data path/clock skew constraint relationships are developed for both the positive clock skew case and the negative clock skew case. From these specific constraint relationships, recommended design procedures were offered to eliminate the deleterious effects of clock skew on the maximum performance and reliable operation of a synchronous digital system. Also, approaches were described for improving synchronous performance by using localized clock skew to equalize delays between local data paths, thereby minimizing the delay of the critical worst case paths.

The results described in this chapter discuss specifically three types of design problems, namely, that in which: 1) L_{max} constrains the design space, 2) f_{clkMAX} constrains the design space, and 3) the design space is unconstrained and a tradeoff must be made between L and f_{clk}. Design equations have been presented which permit a solution to be determined for each type of problem. The solution suggested for the unconstrained problem is the use of design equations which consider the effects of increased latency and increased clock frequency for increasing levels of pipelining. Examples are provided to assist the reader in the application and interpretation of these results.

The goal of this chapter is to assist both the system and circuit designer in developing very high speed complex synchronous digital systems. Systematic approaches are necessary in order to optimally and correctly partition data paths, design clock distribution networks, and select implementing technologies. The results described herein are intended to explore and hopefully clarify these important and related issues.

8. References

[1] E. G. Friedman and J. H. Mulligan, Jr., "Clock Frequency and Latency in Synchronous Digital Systems," *IEEE Transactions on Signal Processing*, Vol. SP-39, No. 4, pp. 930-934, April 1991.

[2] E. G. Friedman and J. H. Mulligan, Jr., "Pipelining of High Performance Synchronous Digital Systems," *International Journal of Electronics*, Volume 70, Number 5, pp. 917-935, May 1991.

[3] M. Hatamian and G. L. Cash, "Parallel Bit-Level Pipelined VLSI Designs for High Speed Signal Processing," *Proceedings of the IEEE*, Vol. 75, No. 9, pp. 1192-1202, September 1987.

[4] E. G. Friedman and S. Powell, "Design and Analysis of a Hierarchical Clock Distribution System for Synchronous Standard Cell/Macrocell VLSI," *IEEE Journal of Solid-State Circuits*, Vol. SC-21, No. 2, pp. 240-246, April 1986.

[5] M. Hatamian, "Understanding Clock Skew in Synchronous Systems," *Concurrent Computations (Algorithms, Architecture, and Technology)*, S. K. Tewksbury, B. W. Dickinson, and S. C. Schwartz (Eds.), pp. 87-96, New York, New York: Plenum Publishing, 1988.

[6] E. G. Friedman, "Clock Distribution Design in VLSI Circuits - an Overview," *Proceedings of IEEE International Symposium on Circuits and Systems*, pp. 1475-1478, May 1993.

130

[7] F. Anceau, "A Synchronous Approach for Clocking VLSI Systems," *IEEE Journal of Solid-State Circuits*, Vol. SC-17, No. 1, pp. 51-56, February 1982.

[8] D. Wann and M. Franklin, "Asynchronous and Clocked Control Structures for VLSI Based Interconnection Networks," *IEEE Transactions on Computers*, Vol. C-32, No. 3, pp. 284-293, March 1983.

[9] S. Dhar, M. Franklin, and D. Wann, "Reduction of Clock Delays in VLSI Structures," *Proceedings of IEEE International Conference on Computer Design*, pp. 778-783, October 1984.

[10] S. H. Unger and C-J. Tan, "Clocking Schemes for High-Speed Digital Systems," *IEEE Transactions on Computers*, Vol. C-35, No. 10, pp. 880-895, October 1986.

[11] H. B. Bakoglu, J. T. Walker, and J. D. Meindl, "A Symmetric Clock-Distribution Tree and Optimized High-Speed Interconnections for Reduced Clock Skew in ULSI and WSI Circuits," *Proceedings of IEEE International Conference on Computer Design*, pp. 118-122, October 1986.

[12] C. V. Gura, "Analysis of Clock Skew in Distributed Resistive-Capacitive Interconnects," University of Illinois, Urbana, Illinois, SRC Technical Report No. T87053, June 1987.

[13] D. Noise, R. Mathews, and J. Newkirk, "A Clocking Discipline for Two-Phase Digital Systems," *Proceedings of International Conference on Circuits and Computers*, pp. 108-111, September 1982.

[14] M. S. McGregor, P. B. Denyer, and A. F. Murray, "A Single-Phase Clocking Scheme for CMOS VLSI," *Proceedings of 1987 Stanford Conference on Advanced Research in VLSI*, pp. 257-271, March 1987.

[15] J. Alves Marques and A. Cunha, "Clocking of VLSI Circuits," *VLSI Architecture*, Randell and Treleaven (Eds.), Englewood Cliffs, New Jersey:Prentice-Hall, 1983, pp. 165-178.

[16] D. Mijuskovic, "Clock Distribution in Application Specific Integrated Circuits," *Microelectronics Journal*, Vol. 18, pp. 15-27, July/August 1987.

[17] K. D. Wagner, "Clock System Design," *IEEE Design & Test of Computers*, pp. 9-27, October 1987.

[18] S. D. Kugelmass and K. Steiglitz, "A Probabilistic Model for Clock Skew," *Proceedings of IEEE International Conference on Systolic Arrays*, pp. 545-554, 1988.

[19] M. Afghahi and C. Svensson, "A Scalable Synchronous System," *Proceedings of International Symposium on Circuits and Systems*, pp. 471-474, May 1988.

[20] M. R. Dagenais and N. C. Rumin, "Automatic Determination of Optimal Clocking Parameters in Synchronous MOS VLSI Circuits," *Proceedings of 1988 Stanford Conference on Advanced Research in VLSI*, pp. 19-33, March 1988.

[21] P. Ramanathan and K. G. Shin, "A Clock Distribution Scheme for Non-Symmetric VLSI Circuits," *Proceedings of International Conference on Computer-Aided Design*, pp. 398-401, November 1989.

[22] E. G. Friedman, *Performance Limitations in Synchronous Digital Systems*, Ph.D. Dissertation, University of California, Irvine, June 1989.

[23] T. Williams and K. Parker, "Design for Testability-A Survey," *Proceedings of the IEEE*, Vol. 71, No. 1, pp. 98-112, January 1983.

[24] H. B. Bakoglu and J. D. Meindl, "Optimal Interconnection Circuits for VLSI," *Proceedings of IEEE International Solid-State Circuits Conference*, pp. 164-165, February 1984.

[25] H. B. Bakoglu and J. D. Meindl, "Optimal Interconnection Circuits for VLSI," *IEEE Transactions on Electron Devices*, Vol. ED-32, No. 5, pp. 903-909, May 1985.

[26] L. W. Cotton, "Circuit Implementation of High-Speed Pipeline Systems," *Proceedings of the Fall Joint Computer Conference*, pp. 489-504, 1965.

[27] P. R. Cappello and K. Steiglitz, "Bit-Level Fixed-Flow Architectures for Signal Processing," *Proceedings of the IEEE International Conference on Circuits and Computers*, pp. 570-573, September 1982.

[28] P. R. Cappello and K. Steiglitz, "Completely-Pipelined Architectures for Digital Signal Processing," *IEEE Transactions on Acoustics, Speech, and Signal Processing*, Vol. ASSP-31, No. 4, pp. 1016-1023, August 1983.

[29] C. E. Leiserson and J. B. Saxe, "Optimizing Synchronous Systems," *Proceedings of 22nd Annual Symposium on Foundations of Computer Science*, pp. 23-26, October 1981.

[30] P. R. Cappello, A. LaPaugh, and K. Steiglitz, "Optimal Choice of Intermediate Latching to Maximize Throughput in VLSI Circuits," *Proceedings of IEEE International Conference on Acoustics, Speech, and Signal Processing*, pp. 935-938, April 1983.

[31] P. R. Cappello, A. LaPaugh, and K. Steiglitz, "Optimal Choice of Intermediate Latching to Maximize Throughput in VLSI Circuits," *IEEE Transactions on Acoustics, Speech, and Signal Processing*, Vol. ASSP-32, No. 1, pp. 28-33, February 1984.

[32] J. R. Jump and S. R. Ahuja, "Effective Pipelining of Digital Systems," *IEEE Transactions on Computers*, Vol. C-27, No. 9, pp. 855-865, September 1978.

[33] M. Hatamian, L. A. Hornak, T. E. Little, S. K. Tewksbury, and P. Franzon, "Fundamental Interconnection Issues," *AT&T Technical Journal*, Volume 66, Issue 4, pp. 13-30, July/August 1987.

[34] K. O. Siomalas and B. A. Bowen, "Synthesis of Efficient Pipelined Architectures for Implementing DSP Operations," *IEEE Transactions on Acoustics, Speech, and Signal Processing*, Vol. ASSP-33, No. 6, pp. 1499-1508, December 1985.

[35] P. Yang and T. N. Trick, "Estimation of Gate Delay in MOS/LSI Circuits," *Proceedings of 22nd Midwest Symposium on Circuits and Systems*, pp. 595-598, June 1979.

[36] R. J. Bayruns, R. L. Johnston, D. L. Fraser, Jr., and S-C. Fang, "Delay Analysis of Si NMOS Gbit/s Logic Circuits," *IEEE Journal of Solid-State Circuits*, Vol. SC-19, No. 5, pp. 755-764, October 1984.

[37] E. T. Lewis, "Optimization of Device Area and Overall Delay for CMOS VLSI Designs," *Proceedings of the IEEE*, Vol. 72, No. 6, pp. 670-689, June 1984.

[38] S. A. Huss and M. Gerbershagen, "Signal Delay Calculation for Integrated CMOS Circuits," *AEU/Archiv fur Elektronik und Ubertragungstechnik*, Vol. 41, pp. 214-222, July/August 1987.

[39] D. Auvergne, D. Deschacht, and M. Robert, "Explicit Formulation of Delays in CMOS VLSI," *Electronic Letters*, Vol. 23, No. 14, pp. 741-742, 2nd July 1987.

[40] P. R. O'Brien, J. L. Wyatt, Jr., T. L. Savarino, and J. M. Pierce, "Fast On-Chip Delay Estimation for Cell-Based Emitter Coupled Logic," Massachusetts Institute of Technology, VLSI Memo no. 88-433, February 1988.

[41] C. Lee and H. Soukup, "An Algorithm for CMOS Timing and Area Optimization," *IEEE Journal of Solid-State Circuits*, Vol. SC-19, No. 5, pp 781-787, October 1984.

[42] H. C. Lin and L. W. Linholm, "An Optimized Output Stage for MOS Integrated Circuits," *IEEE Journal of Solid-State Circuits*, Vol. SC-10, No. 2, pp. 106-109, April 1975.

[43] R. C. Jaeger, "Comments on 'An Optimized Output Stage for MOS Integrated Circuits'," *IEEE Journal of Solid-State Circuits*, Vol. SC-10, No. 3, pp. 185-186, June 1975.

[44] A. Kanuma, "CMOS Circuit Optimization," *Solid-State Electronics*, Vol. 26, No. 1, pp. 47-58, 1983.

[45] H. J. M. Veendrick, "Short-Circuit Dissipation of Static CMOS Circuitry and Its Impact on the Design of Buffer Circuits," *IEEE Journal of Solid-State Circuits*, Vol. SC-19, No. 4, pp. 468-473, August 1984.

[46] L. A. Glasser and L. P. J. Hoyte, "Delay and Power Optimization in VLSI Circuits," *Proceedings of ACM IEEE 21st Design Automation Conference*, pp. 529-535, June 1984.

5

HIGH-SPEED TRANSFORM CODING ARCHITECTURES FOR VIDEO COMMUNICATIONS

C.T. Chiu
Electrical Engr. Dept.
National Chung Cheng University
Chia-Yi, Taiwan

K.J. Ray Liu
Electrical Engr. Dept.
University of Maryland
College Park, MD 20740

INTRODUCTION

The field of signal processing has developed dramatically over the last several decades owing to applications in such diverse fileds as speech, image, and video communication, biomedical engineering, acoustics, consumer electronics, and many others. Discrete sinusoidal transforms such as discrete cosine transform (DCT), discrete sine transform (DST), and discrete Hartley transform (DHT), discrete Fourier transform (DFT), Lopped Orthogonal Transform (LOT), and Complex Lapped Transform (CLT) are powerful tools in many applications of signal processing. Due to the advances in ISDN network and high definition television (HDTV) technology, real-time processing of speech, image and video signals become very desirable. Clearly the high computational rates required in HDTV systems cannot be achieved by general-purpose parallel computers because of severe system overheads. The only way to meet the high computational rates of real-time signal processing is by developing special-purpose architectures which exploit the regularity, recursiveness, and locality of the signal processing algorithms. We focus on developing real time VLSI algorithms and architectures for discrete sinusoidal transforms for video applications.

A "time-recursive" approach is proposed to perform transform coding on a real time basis. The transformed data are updated according to a recursive formula, whenever new data arrive. Based on this concept, new unified parallel lattice architectures for the DCT, DST, DHT, DFT, LOT, CLT are derived. Here "unified" architecture means that different transforms can be computed using the same structure. We reduce the number of multipliers from $\frac{N}{2} * \ln N$ to $6N - 8$ for 1-D DCT. Moreover, the resulting architectures are regular, modular, locally-connected and suitable for VLSI implementation.

From the speed point of view, this unified architecture can obtain the the transform results immediately whenever a new datum arrives. Therefore, the system throughput rate is highly increased and is better than others for matching the high speed requirement of video communication systems. Moreover, there is no constraint on the size of the image.

Data processed in image and video signal processing are two-dimensional (2-D) information. The drawback of the conventional 2-D transforms is the delay time due to a operation called "transposition". We derive a new time-recursive parallel 2-D DCT structure which can eliminate the transposition time. The system is fully-pipelined with throughput rate N clock cycles for $N \times N$ successive input data frame.

The VLSI implementation of the lattice module based on the distributed arithmetic is also described. This is the first chip that can dually generate the DCT and DST simultaneously. It has been fabricated using $2\mu m$ double-metal CMOS technology and has been tested to be fully functional with a throughput rate 14.5-MHz and a data processing rate of 116Mb/s.

To summarize, this chapter addresses the problem of developing efficient VLSI algorithms and architectures for discrete sinusoidal transforms in real-time applications for video communication systems. We propose unified time-recursive algorithms, architectures, and implement the proposed DCT/DST structure into a IC chip.

PARALLEL LATTICE STRUCTURES

In this section, the unified time-recursive lattice structures that can be used for the discrete orthogonal transforms mentioned above is proposed. We consider the orthogonal transforms from a time-recursive point of view instead of the whole block of data.

Dual Generation of DCT and DST

We will show an efficient implementation of the DCT from the time-recursive point of view. Focusing on the sequence instead of the block of input data, we can obtain not only the time-recursive relation between the DCT of two successive data sequences, but also a fundamental relation between DCT and DST. In the following, the time-recursive relation for the DCT will be considered first.

The one-dimensional (1-D) DCT of a sequential input data starting from $x(t)$ and ending with $x(t + N - 1)$ is defined as

$$X_c(k, t) = \frac{2C(k)}{N} \sum_{n=t}^{t+N-1} x(n) \cos\left[\frac{\pi[2(n - t) + 1]k}{2N}\right]$$
$$k = 0, 1, ..., N - 1, \tag{1}$$

where

$$C(k) = \begin{cases} \frac{1}{\sqrt{2}} & \text{if } k = 0 \text{ or } k = N, \\ 1 & \text{otherwise.} \end{cases}$$

Here the time index t in $X_c(k, t)$ denotes that the transform starts from $x(t)$. Since the function $C(k)$ has a different value only when $k = 0$, we can consider those cases that $C(k)$'s equal one (*i.e.* $k = 1, 2, ..., N - 1$.) first and re-examine the case for $k = 0$ later on. In transmission systems data arrive seriesly, therefore we are interested in the the 1-D DCT of the next input data vector $[x(t+1), x(t+2), ..., x(t+N)]$. From the definition, it is given by

$$X_c(k, t+1) = \frac{2}{N} \sum_{n=t+1}^{t+N} x(n) \cos \left[\frac{\pi[2(n-t-1)+1]k}{2N} \right]. \tag{2}$$

This can be rewritten as

$$X_c(k, t+1) = \overline{X}_c(k, t+1) \cos \left(\frac{\pi k}{N} \right) + \overline{X}_s(k, t+1) \sin \left(\frac{\pi k}{N} \right), \tag{3}$$

where

$$\overline{X}_c(k, t+1) = \frac{2}{N} \sum_{n=t+1}^{t+N} x(n) \cos \left[\frac{\pi[2(n-t)+1]k}{2N} \right], \tag{4}$$

and

$$\overline{X}_s(k, t+1) = \frac{2}{N} \sum_{n=t+1}^{t+N} x(n) \sin \left[\frac{\pi[2(n-t)+1]k}{2N} \right]. \tag{5}$$

As we can see, a DST-like term $\overline{X}_s(k, t+1)$ appears in (5). This motivates us to investigate the time-recursive DST.

There are several definitions for the DST. Here we prefer the definition proposed by Wang in [19]. The 1-D DST of a data vector $[x(t), x(t+1), ..., x(t+N-1)]$ is defined as

$$X_s(k, t) = \frac{2C(k)}{N} \sum_{n=t}^{t+N-1} x(n) \sin \left[\frac{\pi[2(n-t)+1]k}{2N} \right],$$

$$k = 1, ..., N. \tag{6}$$

Note that the range of k is from 1 to N. Again, we consider those cases that $D(k)$'s equal one first, $i.e.$

$$X_s(k, t) = \frac{2}{N} \sum_{n=t}^{t+N-1} x(n) \sin \left[\frac{\pi[2(n-t)+1]k}{2N} \right]. \tag{7}$$

The DST of the time update sequence $[x(t+1), x(t+2), ..., x(t+N)]$ is given by

$$X_s(k, t+1) = \frac{2}{N} \sum_{n=t+1}^{t+N} x(n) \sin \left[\frac{\pi[2(n-t-1)+1]k}{2N} \right]$$

$$= \overline{X}_s(k, t+1) \cos \left(\frac{\pi k}{N} \right) - \overline{X}_c(k, t+1) \sin \left(\frac{\pi k}{N} \right). \tag{8}$$

Here the terms $\overline{X}_s(k, t+1)$ and $\overline{X}_c(k, t+1)$ that are used in (3) to generate $X_c(k, t+1)$ appear in the equation of the new DST transform $X_s(k, t+1)$ again. This suggests that the DCT and DST can be dually generated from each other.

138

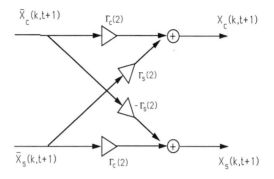

$\Gamma_c(n)=\cos(\pi kn/2N),$　　$\Gamma_s(n)=\sin(\pi kn/2N)$

Figure 1: The lattice module.

The Lattice Structures

From (3) and (8), it is noted that the new DCT and DST transforms $X_c(k,t+1)$ and $X_s(k,t+1)$, can be obtained from $\overline{X}_c(k,t+1)$ and $\overline{X}_s(k,t+1)$ in the lattice form as shown in Fig. 1. The next step is to update $\overline{X}_c(k,t+1)$ and $\overline{X}_s(k,t+1)$ from the previous transforms $X_c(k,t)$ and $X_s(k,t)$. We notice that $X_c(k,t)$ and $\overline{X}_c(k,t+1)$ have similar terms except the old datum $x(t)$ and the incoming new datum $x(t+N)$. Therefore $\overline{X}_c(k,t+1)$ and $\overline{X}_s(k,t+1)$ can be obtained by deleting the term associated with the old datum $x(t)$ a nd updating the new datum $x(t+N)$ as

$$\overline{X}_c(k,t+1) = X_c(k,t) + \left[-x(t) + (-1)^k x(t+N)\right] \left(\frac{2}{N}\right) \cos\left(\frac{\pi k}{2N}\right), \quad (9)$$

and

$$\overline{X}_s(k,t+1) = X_s(k,t) + \left[-x(t) + (-1)^k x(t+N)\right] \left(\frac{2}{N}\right) \sin\left(\frac{\pi k}{2N}\right). \quad (10)$$

From (3), (8), (9), and (10), the new transforms $X_c(k,t+1)$ and $X_s(k,t+1)$ can be calculated from the previous transforms $X_c(k,t)$ and $X_s(k,t)$ by adding the effect of input signal samples $x(t)$ and $x(t+N)$. This demonstrates that the DCT and DST can be dually generated from each other in a recursive way.

The complete time-recursive lattice modules for ($k = 0, 1, 2, .., N-1$.) are shown in Fig. 2. It consists of a $N+1$ shift register and a normalized digital filter performing the plane rotation. The multiplications in the plane rotation can be reduced to addition and substration for $k = 0$ in the DCT and $k = N$ in the DST respectively. The following illustrates how this dually generated DCT and DST lattice structure works to obtain the DCT and DST with length N of a series of input data $[x(t), x(t+1), .., x(t+N-1), x(t+N), ...]$ for a specific k. The initial values of the transformed signals $X_c(k,t-1)$ and $X_s(k,t-1)$ are set to zero; so are the initial values in the shift register in the front of the lattice

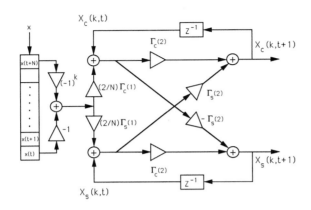

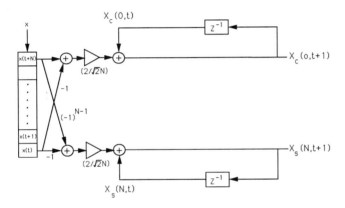

Figure 2: The lattice structure for the DCT and DST with coefficients $C(k)$'s and $D(k)$'s, $k = 0, 1, 2, ..., N-1, N$.

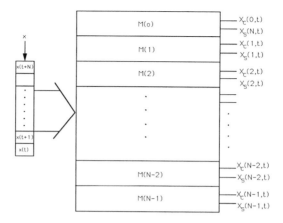

Figure 3: The parallel lattice structure for the DCT and DST.

module. The input sequence $[x(t), x(t + 1), ...]$ shifts sequentially into the shift register as shown in Fig. 2. Then the output signals $X_c(k, t)$ and $X_s(k, t)$, $k = 0, 1, ..., N - 1, N$, are updated recursively. After the input datum $x(t + N - 1)$ shifts into the shift register, the DCT and DST of the input data vector $[x(t), x(t+1), ..., x(t+N-1)]$ are obtained at the output for this index k. It takes N clock cycles to get the $X_c(k, t)$ and $X_s(k, t)$ of the input vector $[x(t), x(t + 1), ..., x(t + N - 1)]$. Since there are N different values for k, the total computational time to obtain all the transformed data is N^2 clock cycles, if only one lattice module is used. In this case, the delay time and throughput are the same N^2 clock cycles.

A parallel lattice array consists of N lattice modules can be used for parallel computations and it improves the computational speed drastically as shown in Fig. 3. Here we have seen that the transform domain data $X(k, t)$ have been decomposed into N disjoint components that have the same lattice modules with different multiplier coefficients in them. In this case the total computational delay time decreases to N clock cycle. It is important to notice that when the next input datum $x(t + N)$ arrives, the transformed data of the input data vector $[x(t+1), x(t+2), ..., x(t+N)]$ can be obtained immediately. Likewise, it takes only one clock cycle to generate the transformed data of subsequent inputs. That is, the latency and throughput of this parallel system are N and 1 respectively.

Inverse Transforms

According to the definition of the DCT in (1), the IDCT for the transform domain sequence $[X(t), X(t + 1), ..., X(t + N - 1)]$ is

$$x_c(n, t) = \sum_{k=t}^{t+N-1} C(k - t)X(k) \cos \left[\frac{\pi(2n + 1)(k - t)}{2N} \right], \qquad (11)$$

The coefficients $C(k)$'s are given in (1) and n is from zero to $N - 1$.. From the time-recursive point of view, the IDCT of the new sequence $[X(t + 1), X(t + 2), ..., X(t + N)]$ can be expressed as

$$x_c(n, t + 1) = \sum_{k=t+1}^{t+N} C(k - t - 1)X(k) \cos \left[\frac{\pi(2n + 1)(k - t - 1)}{2N} \right]. \quad (12)$$

Similar to the previous sections, we can decompose (12) into

$$x_c(n, t + 1) = \overline{x}_c(n, t + 1) \cos \left[\frac{\pi(2n + 1)}{2N} \right] + \overline{x}_{as}(n, t + 1) \sin \left[\frac{\pi(2n + 1)}{2N} \right], \quad (13)$$

where

$$\overline{x}_c(n, t + 1) = \sum_{k=t+1}^{t+N} C(k - t - 1)X(k) \cos \left[\frac{\pi(2n + 1)(k - t)}{2N} \right], \quad (14)$$

and

$$\overline{x}_{as}(n, t + 1) = \sum_{k=t+1}^{t+N} C(k - t - 1)X(k) \sin \left[\frac{\pi(2n + 1)(k - t)}{2N} \right]. \quad (15)$$

In order to be a dually generated pair of the IDCT given in (11), we define the auxiliary inverse discrete sine transform (AIDST) as

$$x_{as}(n, t) = \sum_{k=t}^{t+N-1} C(k - t)X(k) \sin \left[\frac{\pi(2n + 1)(k - t)}{2N} \right], \quad (16)$$

Although this definition utilizes the same sine functions as the transform kernel, it is not the inverse transform of the DST. We call this the AIDST. By using the trigonometric function expansions, $x_{as}(n, t + 1)$ becomes

$$x_{as}(n, t + 1) = \overline{x}_{as}(n, t + 1) \cos \left[\frac{\pi(2n + 1)}{2N} \right] - \overline{x}_c(n, t + 1) \sin \left[\frac{\pi(2n + 1)}{2N} \right]. \quad (17)$$

Lattice Structure for IDCT

Combining (13) and (17), we observe that the IDCT and the AIDST can be generated in exactly the same way as the dual generation of the DCT and DST. Therefore, the lattice structure in Fig.1 can be applied here except that the coefficients must be modified. Since the coefficients $C(k)$'s are inside the expression in the inverse transform, the relation between $x_c(n, t)$ and $\overline{x}_c(n, t + 1)$ will be different from what we have in the DCT. Equations (11) and (14) as well as (15) and (16) have the same terms for $k \in \{t + 2, t + 3, ..., t + N - 1\}$. After adding the effects of the terms for $k = t$ and $k = t + 1$, we obtain

$$\overline{x}_c(n, t + 1) = x_c(n, t) - \frac{1}{\sqrt{2}}X(t) + \left(\frac{1}{\sqrt{2}} - 1 \right) \cos \left[\frac{\pi(2n + 1)}{2N} \right] X(t + 1), \quad (18)$$

142

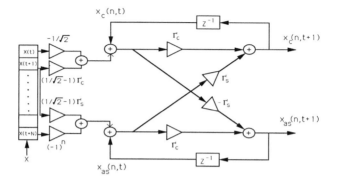

$r'_c(n)=\cos(\pi(2n+1)/2N), \quad r'_s(n)=\sin(\pi(2n+1)/2N)$

Figure 4: The lattice structure for the IDCT and AIDST.

and

$$\bar{x}_{as}(n,t+1) = x_{as}(n,t)+(-1)^n X(t+N)+\left(\frac{1}{\sqrt{2}}-1\right)\sin\left[\frac{\pi(2n+1)}{2N}\right]X(t+1).$$
(19)

The complete lattice module for the IDCT and AIDST is shown in Fig.4. This IDCT lattice structure has the same lattice module as that of the DCT except for the input stage where one more adder and one more multiplier are required. The procedure to calculate the inverse transformed data is the same. Therefore, this IDCT lattice structure has the same advantages as that of the DCT. To obtain the inverse transform in parallel, we need N such IDCT lattice modules where $7N$ multipliers and $6N$ adders are required. Again, we see that the numbers of adders and multipliers are linear functions of N. Here we should notice that to obtain the inverse transform of the original input data sequence, for example, $[x(0), x(1), x(2), ..., x(N-1)]$ and $[x(N), x(N+1), ..., x(2N-1)]$, it is sufficient only to send the transformed data corresponding to these two blocks, i.e., $[X(0), X(1), ..., X(N-1)]$ and $[X(N), X(N+1), ..., X(2N-1)]$ respectively, although we have all the intermediate transformed data. Then by applying the time-recursive algorithm mentioned above, we obtain the original data after $X(N-1)$ and $X(2N-1)$ arrive, the intermediate data obtained by the inverse transform are redundant.

Unified Lattice Structures

The definition of DHT is given by[2]

$$X_h(k,t) = \frac{1}{\sqrt{N}}\sum_{n=t}^{t+N-1} x(n)cas\left(2(n-t)\frac{\pi k}{N}\right)$$
$$k = 0, 1, ..., N-1,$$
(20)

where $cas\theta \triangleq \cos\theta + \sin\theta$.

The Discrete Fourier Transform (DFT) of N samples $[x(t), x(t+1), ..., x(t+$

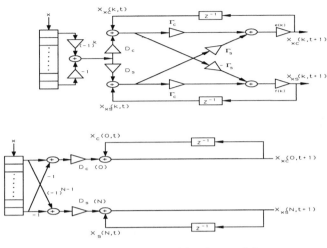

Figure 5: The universal lattice module.

$N - 1)$] is defined as[27]

$$X_f(k,t) = \frac{1}{\sqrt{N}} \sum_{n=t}^{t+N-1} x(n) \exp\{-j2(n-t)\frac{\pi k}{N}\}, \quad k = 0, 1, \ldots, N-1.$$

(21)

The DHT uses real expressions $\cos(\frac{2\pi k(n-t)}{N}) + \sin(\frac{2\pi k(n-t)}{N})$ as the transform kernel, while discrete Fourier transform (DFT) uses the complex exponential expression $\exp(\frac{j2\pi k(n-t)}{N})$ as the transform kernel. Because the kernel of the DHT is a summation of cosine and sine terms, we can separate them into a combination of a DCT-like and a DST-like transforms as follows:

$$X_h(k,t) = \acute{X}_c(k,t) + \acute{X}_s(k,t),$$

(22)

where

$$\acute{X}_c(k,t) = \frac{1}{\sqrt{N}} \sum_{n=t}^{t+N-1} x(n) \left[\cos\left(\frac{2\pi k(n-t)}{N}\right)\right],$$

(23)

and

$$\acute{X}_s(k,t) = \frac{1}{\sqrt{N}} \sum_{n=t}^{t+N-1} x(n) \left[\sin\left(\frac{2\pi k(n-t)}{N}\right)\right].$$

(24)

The $\acute{X}_c(k,t)$ is the so-called DCT-I and the $\acute{X}_s(k,t)$ is the DST-I that are defined by Yip and Rao in [20]. The $\acute{X}_c(k,t)$ and $\acute{X}_s(k,t)$ can be dually generated from each other with the corresponding coefficients listed in Table 1. Therefore, the DCT-I and DST-I are dual pairs. From the lattice structure shown in Fig. 5, the DHT are obtained by adding the the dual pairs $\acute{X}_c(k,t)$ and $\acute{X}_s(k,t)$. The real part of the DFT is $\acute{X}_c(k,t)$ and and the imaginary part is $\acute{X}_s(k,t)$.

	DCT/DST	DHT/DFT	LOT/CLT
Γ_c	$\cos(\pi k/N)$	$\cos(2\pi k/N)$	$\cos(\pi/2N)$
Γ_s	$\sin(\pi k/N)$	$\sin(2\pi k/N)$	$\sin(\pi/2N)$
D_c	$C(k)\sqrt{\frac{2}{N}}\cos(\pi k/2N)$	$\sqrt{\frac{1}{N}}$	$\sqrt{\frac{1}{N}}(-1)^k j \cdot$ $\exp-j\theta_k \sin(\pi/4N)$
D_s	$C(k)\sqrt{\frac{2}{N}}\sin(\pi k/2N)$	0	$\sqrt{\frac{1}{N}}(-1)^k j \cdot$ $\exp-j\theta_k \cos(\pi/4N)$
$e(k)$	1	1	$\exp j2\theta_k$
$f(k)$	1	1	$\exp j2\theta_k$

Table 1: Coefficients of the Lattice structure for the DXT

The Complex Lapped Transform (CLT) [29] of $2N$ samples $[x(t - N + \frac{1}{2}), x(t - N + \frac{3}{2}), \ldots, x(x + N - \frac{1}{2})]$ is defined as

$$X_{clt}(k,t) = \frac{1}{\sqrt{N}} \sum_{n=t-(N-\frac{1}{2})}^{t+(N-\frac{1}{2})} x(n)\exp\{-j\frac{(2k+1)(n-t)\pi}{2N}\} \cos\frac{(n-t)\pi}{2N},$$

$$k = 0, 1, \ldots, N - 1. \qquad (25)$$

The Lapped Orthogonal Transform (LOT) [29] of $2N$ samples $[x(t - N + \frac{1}{2}), x(t - N + \frac{3}{2}), \ldots, x(t + N - \frac{1}{2})]$ is defined as

$$X_{lot}(k,t) = \begin{cases} \sqrt{\frac{2}{N}} \sum_{n=t-(N-\frac{1}{2})}^{t+(N-\frac{1}{2})} x(n) \cos\frac{(2k+1)(n-t)\pi}{2N} \cos\frac{(n-t)\pi}{2N} + \alpha_k, \\ \qquad k = 0, 2, \ldots, (N-2), \text{even part of the CLT} \\ \sqrt{\frac{2}{N}} \sum_{n=t-(N-\frac{1}{2})}^{t+(N-\frac{1}{2})} x(n) \sin\frac{(2k+1)(n-t)\pi}{2N} \cos\frac{(n-t)\pi}{2N} + \beta_{nk}, \\ \qquad k = 1, 3, \ldots, (N-1), \text{odd part of the CLT} \end{cases}$$

$$(26)$$

where $\alpha_k = \beta_{nk} = 0$, except for $\alpha_0 = -(\sqrt{2}-1)/(2\sqrt{2})$, and $\beta_{n(N-1)} = (-1)^{n+\frac{1}{2}}\alpha_0$.

Since the LOT is obtained from the even and odd value of k, we focus on the discussion of the dual generation for the CLT only. Define an Auxiliary Complex Lapped Transform (ACLT) of $2N$ samples $[x(t - N + \frac{1}{2}), x(t - N + \frac{3}{2}), \ldots, x(x + N - \frac{1}{2})]$ as

$$X_{clt}(k,t) = \frac{1}{\sqrt{N}} \sum_{n=t-(N-\frac{1}{2})}^{t+(N-\frac{1}{2})} x(n)\exp\{-j\frac{(2k+1)(n-t)\pi}{2N}\} \sin\frac{(n-t)\pi}{2N},$$

$$k = 0, 1, \ldots, N - 1. \qquad (27)$$

Then, the CLT and ACLT can be dually generated from the DCT and DST with the corresponding coefficients listed in Table 1. All the transforms mentioned above can be realized by a lattice structure as shown in Fig. 5. This lattice structure is a modified normal form digital filter. Table 1 lists the coefficients in the unified lattice structure for different transforms. Here θ_k associated with the LOT/CLT equals $\frac{(2k+1)\pi}{4N}$.

	LiuChiu	chen [14] et. al.	Lee[2]	Hou[5]
No. of Multipliers	$6N - 4$	$N \ln(N) -3N/2 + 4$	$(N/2) \ln(N)$	$N - 1$
latency	N	$N/2$	$\frac{1}{2}[\ln(N)* (\ln(N) - 1)]$	$O(3N/2)$
limitation on N	no	powerof 2	power of 2	power of 2
commun.	local	global	global	global
I/O	SIPO	PIPO	PIPO	SIPO

Table 2: Comparison of different DCT algorithms

Comparisons of Architectures

From the previous discussions, we see that the proposed unified parallel lattice structures have many attractive features. There are no constraints on the transform size N. It dually generates the two discrete transforms DCT and DST simultaneously. Since it produces the transformed data of subsequent input vector every c input data such as communication systems. Further, the structure is regular, modular, and without global communication. As a consequence, it is suitable for VLSI implementation.

Here, we would like to compare our lattice structures of the DCT and DST with those proposed in [7, 8, 4]. The architecture in [7] uses the matrix factorization method which is a representative of fast algorithms. In [8], an improved fast structure with fewer multipliers is proposed. Hou's architecture in [4] uses recursive computations to generate the higher order DCT from the lower order one. The characteristics of these structures are discussed in the introduction. A comparison regarding their inherent properties is listed in Table 2. The lattice architecture with six multipliers in the module as shown in Fig. 2 is called Liu-Chiu structure. The structure in Liu-Chiu1 has $6N - 4$ multipliers, $5N - 1$ adders, and the latency is N. There are $4N$ multipliers, $5N - 1$ adders, and the latency is $2N$ in the structure of Liu-Chiu2. The number of multipliers is reduced by the order $2N$ in the expense of doubling the latency and the data flow becoming $SISO$. The Liu-Chiu3 architecture has $5N$ multipliers and $7N$ adders and the latency is N clock cycles. From these Tables, it is noted that the number of multipliers in our architectures is higher than that of others when N is small. This is due to the dual generation of two transforms structure which is compatible with Lee's. Since the numbers of multipliers and adders of our structures are on the order N, our algorithms have fewer multipliers and adders than those proposed in [7, 8]. Although Hou's algorithm has the fewest multipliers, his architecture needs global communications and the design complexity is more than of other structures can not start until all of the data in the block arrive.

A comparison for our DHT structure based on the lattice module and different DHT algorithms [21, 34] is listed in Table 3. The architecture in [21], a representative fast algorithm, depends on the existing FFT method. Chaitali-JaJa's algorithm in [34] decomposes the transform size N into mutually prime numbers and implements them in a systolic manner. Their structure needs extra registers and the latency is higher than others.

	Liu-Chiu	Sorenson[23]	Chaitali-JaJa[18]
No. of Multipliers	$4N$	$N\ln(N)$ $-3N+4$	$N1+N2$
Adders	$5N-2$	$3N\ln(N)$ $-3N+4$	$4N$ $+\sqrt{N1}$
latency	N	$N\ln(N)$	$N1+N2$
limitation on N	no	power of 2	$N=N1*N2$, $N1, N2$ are m.p.
commun.	local	global	local
I/O	SIPO	PIPO	SISO

Table 3: Comparison of different DHT algorithms,*m.p. means "mutual prime".

TWO-DIMENSIONAL DCT LATTICE STRUCTURES

The 2-D DCT has been widely recognized as the most effective technique in image data compression, especially for high-speed video processing applications, such as "HDTV." In this section, a new fully-pipelined architecture to compute the 2-D DCT from a frame-recursive point of view is proposed.

Dual Generation of 2-D DCT and DSCT

By focusing directly on the 2-D transformed signal and applying the frame recursive approach, we can derive not only the frame recursive relation of two successive frames of the 2-D DCT, but also the dual generation properties between the 2-D DCT and 2-D discrete sine-cosine transform (DSCT). Here the DSCT serves as an auxiliary transform which supports the time-recursive computations of the 2-D DCT.

Frame-Recursive 2-D Discrete Cosine Transform

The $N \times N$ 2-D DCT $\{X_c(k,l,t) : k,l = 0, 1, ..., N-1.\}$ of an $N \times N$ 2-D data sequence $\{x(m,n) : m = t, t+1, ..., t+N-1; n = 0, 1, ..., N-1.\}$ is defined as

$$X_c(k,l,t) = \frac{4}{N^2}C(k)C(l)\sum_{m=t}^{t+N-1}\sum_{n=0}^{N-1} x(m,n)\cos\left[\frac{\pi[2(m-t)+1]k}{2N}\right] \tag{28}$$
$$\cdot \cos\left[\frac{\pi(2n+1)l}{2N}\right].$$

Here the time index t in $X_c(k,l,t)$ denotes that the transform starting from the t'th row of the 2-D data sequence $\{x(m,n) : m = 0, 1, 2, ...; n = 0, 1, ..., N-1.\}$ as shown in Fig. 6. In the following, we call $X_c(k,l,t)$ the 2-D DCT of the t'th frame of the 2-D data sequence $x(m,n)$. To derive the time-recursive relations between the successive data frames, let us start by considering the 2-D DCT of the t'th frame data sequence,

$$X_c(k,l,t) = \frac{4}{N^2}C(k)C(l)\sum_{m=t}^{t+N-1}\sum_{n=0}^{N-1} x(m,n)\cos\left[\frac{\pi(2(m-t)+1)k}{2N}\right] \tag{29}$$

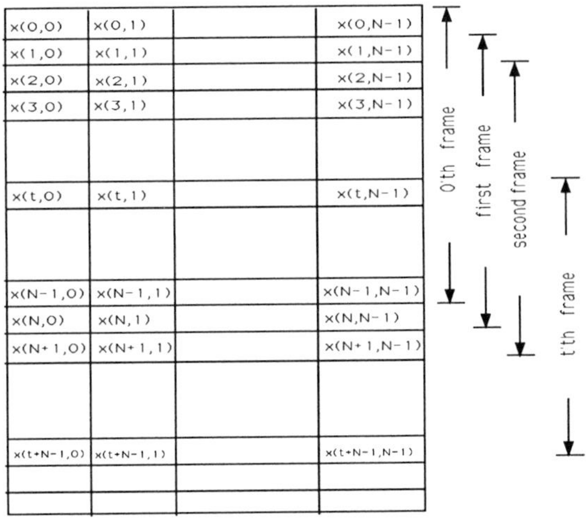

x(0,0)	x(0,1)		x(0,N-1)
x(1,0)	x(1,1)		x(1,N-1)
x(2,0)	x(2,1)		x(2,N-1)
x(3,0)	x(3,1)		x(3,N-1)
x(t,0)	x(t,1)		x(t,N-1)
x(N-1,0)	x(N-1,1)		x(N-1,N-1)
x(N,0)	x(N,1)		x(N,N-1)
x(N+1,0)	x(N+1,1)		x(N+1,N-1)
x(t+N-1,0)	x(t+N-1,1)		x(t+N-1,N-1)

(right side labels: 0th frame, first frame, second frame, t th frame)

Figure 6: The 2-D successive data frame.

$$\cdot \cos\left[\frac{\pi(2n+1)l}{2N}\right].$$

Instead of focusing on $X_c(k,l,t)$ and utilizing various techniques to reduce the computational complexity. We will consider the 2-D DCT sequence of the $(t+1)$'th frame, which is

$$X_c(k,l,t+1) = \frac{4}{N^2}C(k)C(l)\sum_{m=t+1}^{t+N}\sum_{n=0}^{N-1}x(m,n)\cos\left[\frac{\pi[2(m-t-1)+1]k}{2N}\right]$$
$$\cdot \cos\left[\frac{\pi(2n+1)l}{2N}\right]. \tag{30}$$

By using trigonometric function expansions on $\cos\left[\frac{\pi[2(m-t-1)+1]k}{2N}\right]$, (30) can be rewritten as

$$X_c(k,l,t+1) = \overline{X}_c\cos\left(\frac{\pi k}{N}\right) + \overline{X}_{sc}\sin\left(\frac{\pi k}{N}\right), \tag{31}$$

where

$$\overline{X}_c = \frac{4}{N^2}C(k)C(l)\sum_{m=t+1}^{t+N}\sum_{n=0}^{N-1}x(m,n)\cos\left[\frac{\pi[2(m-t)+1]k}{2N}\right]\cos\left[\frac{\pi(2n+1)l}{2N}\right], \tag{32}$$

and

$$\overline{X}_{sc} = \frac{4}{N^2}C(k)C(l)\sum_{m=t+1}^{t+N}\sum_{n=0}^{N-1}x(m,n)\sin\left[\frac{\pi[2(m-t)+1]k}{2N}\right]\cos\left[\frac{\pi(2n+1)l}{2N}\right]. \tag{33}$$

We can view the term $\sin\left[\frac{\pi[2(m-t)+1]k}{2N}\right]\cos\left[\frac{\pi(2n+1)l}{2N}\right]$ in (33) as a new transform kernel, and define a $N \times N$ 2-D discrete sine-cosine transform (DSCT) sequence $\{X_{sc}(k,l,t) : k = 1, 2, ..., N; l = 0, 1, ..., N-1.\}$ of the 2-D data sequence $\{x(m,n) : m = t, t+1, ..., N+t-1; n = 0, 1, ..., N-1.\}$ as

$$X_{sc}(k,l,t) = \frac{4}{N^2}C(k)C(l)\sum_{m=t}^{N+t-1}\sum_{n=0}^{N-1}x(m,n)\cos\left[\frac{\pi(2n+1)l}{2N}\right] \tag{34}$$
$$\cdot\sin\left[\frac{\pi[2(m-t)+1]k}{2N}\right].$$

Here we extend the definition of $C(k)$ to $C(N)$ and define $C(N) = \frac{1}{\sqrt{2}}$. Similarly, we are interested in the 2-D DSCT of the $(t+1)$'s frame. According to the definition, it is

$$X_{sc}(k,l,t+1) = \overline{X}_{sc}\cos\left(\frac{\pi k}{N}\right) - \overline{X}_c\sin\left(\frac{\pi k}{N}\right), \tag{35}$$

where the terms $\overline{X}_c$ and $\overline{X}_{sc}$ used in (31) to generate $X_c(k,l,t+1)$ appear again. This suggests that the 2-D DCT and 2-D DSCT can be dually generated from each other.

Lattice structures for frame-recursive 2D-DCT

We will show in this section that 2D-DCT can be generated by using two lattice arrays. From (31) and (35), it is noted that the new 2-D DCT and DSCT transforms can be obtained from the intermediate values $\overline{X}_c$ and $\overline{X}_{sc}$ in a lattice form as shown in Fig. 2 except removing the shift register in the front.

A similar relation also exists in the dual generations of the 1-D DCT and 1-D DST [10]. The intermediate data $\overline{X}_c$ and $\overline{X}_{sc}$ differ from the original signal $X_c(k,l,t)$ and $X_{sc}(k,l,t)$ only in the t'th row and the $(t+N)$'th row of the input data frames. So, the intermediate data $\overline{X}_c$ and $\overline{X}_{sc}$ can be obtained from $X_c(k,l,t)$ and $X_{sc}(k,l,t)$ by removing the t'th row of the transformed data and updating the $(t+N)$'th row of the transformed data. That is,

$$\overline{X}_c = X_c(k,l,t) + \delta_c(k,l,t)\frac{2}{N}C(k)\cos\left(\frac{\pi k}{2N}\right), \tag{36}$$

and

$$\overline{X}_{sc} = X_{sc}(k,l,t) + \delta_c(k,l,t)\frac{2}{N}C(k)\sin\left(\frac{\pi k}{2N}\right), \tag{37}$$

where

$$\delta_c(k,l,t) = \frac{2}{N}C(l)\sum_{n=0}^{N-1}\left[(-1)^k x(t+N,n) - x(t,n)\right]\cos\left[\frac{\pi(2n+1)l}{2N}\right]. \tag{38}$$

By substituting $\overline{X}_c$ and $\overline{X}_{sc}$ in (36) and (37) into the updated transformed signal $X_c(k,l,t+1)$ and $X_{sc}(k,l,t+1)$ in (31) and (35), the relation between the updated transformed signal and previous transformed signal for

$k = 1, 2, .., N - 1$ are represented in a lattice form as shown in the upper part of Fig. 2. Since the multiplications can be reduced to addition and subtraction for $k = 0$ in the 2-D DCT and $k = N$ in the DSCT respectively, these two cases can be simplified as shown in the lower part of Fig. 2. What is worth noticing is that $\delta_c(k, l, t)$ in (38) is the 1-D DCT of the data vector which is the difference between the parity of the t'th row and $(t + N)$'th row of the 2-D input sequence. As described in [10], $\delta_c(k, l, t)$ can be generated in a lattice form as shown in Fig. 2.

We call the lattice array with one delay in the feedback path as lattice array I (LAI) and that one with N delays in the feedback as lattice array II $(LAII)$. Comparing these two structures, we observe that these two lattice modules have the same architecture except that LAI feedbacks the outputs directly through a delay stage to add with the inputs.

The 2-D DCT and DSCT are produced by applying input data frames to LAIs which generate the $\delta_c(k, l, t)$. After obtaining the $\delta_c(k, l, t)$, the updated transformed signal can be obtained recursively by feeding $\delta_c(k, l, t)$ into the lattice array II. We observe that the 2-D DCT can be obtained by using two 1-D DCT lattice arrays. It will be sho approach is fully-pipelined and no transposition is required. This is because that by using the frame-recursive approach, we start from the transformed 2-D DCT directly and avoid calculating the 2-D DCT indirectly from the 1-D DCT. Our architectures are efficient since it is a direct 2-D approach. This method can also obtain the 2-D DCT and 2-D DSCT simultaneously. In contrast to processing the input 2-D data sequence by rows, the input data can be updated by columns. In this case, 2-D DCT and 2-D discrete cosine-sine transform (DCST) are dually generated, and all other results are similar and can be easily derived.

Architectures of Frame-Recursive Lattice 2D-DCT and 2-D DSCT

We will discuss two architectures, the moving frame 2-D DCT architecture and the block 2-D DCT architecture. The moving frame 2-D DCT architecture is used to calculate the 2-D DCT of sequential frames. For example, the 2-D DCTs of the 0'th frame, first frame, second frame and so on. The block 2-D DCT architecture computes the 2-D DCT of an $N \times N$ input data matrix for every N frames, $i.e.$, the 0'th frame, the N'th frame, the $2N$'th frame and so on.

The Moving Frame 2-D DCT Architectures

The moving frame 2-D DCT architectures generate the 2-D DCT of successive input frames. From the frame-recursive concept derived in section 3.1, the 2-D DCT recursive lattice structures can be constructed according to (31), (35), and (38). Although the intermediate values $\delta_c(k, l, t)$ in (38) are functions of both k and l, it is noted that the effect due to the index k is equivalent to sign changes in the input data sequences. Using this property, we will show two approaches, the pre-matrix method and the post-matrix method, to implement the moving frame architectures. The pre-matrix method will be discussed first.

The Pre-Matix Method

In this method, the intermediate values of $\delta_c(k, l, t)$ are realized directly from (38). As we can see, the index k in (38) affect only the sign of the new input data sequence. Thus, there are only two possible input sequence combinations: $\{ x(t+N,n) - x(t,n)\}$ and $\{ -x(t+N,n) - x(t,n)\}$. The resulting pre-matrix moving-frame 2-D DCT architecture is shown in Fig. 7 which includes a circular shift matrix I, two adders, two LAIs, one $LAII$, and two circular shift arrays and shift register arrays. Except for the LAI, $LAII$, and adders, all other components are shift registers. We will describe the functions of every blocks first, then demonstrate how the system works.

The circular shift matrix I ($CSMI$) is an $(N+1) \times N$ shift registers connected as shown in the upper part of Fig. 8. When a new input datum $x(m,n)$ arrives every clock cycle, all the data are shifted in the way as indicated in Fig. 8. Both of the first elements in the t'th row and $(t+N)$'th row are sent to the adders for summation and subtraction as shown in Fig. 7. The pre-matrix architecture contains two LAIs which includes N lattice modules. The upper and lower LAIs execute the 1-D DCT on the rows of the 2-D input data for the even and odd transformed components k respectively. Because the length of the input vector is N and only the discrete cosine transformed data are needed, the 1-D DCT transformed data $\delta_c(k,l,t)$ generated by the LAIs are sent out every N clock cycles [10]. Due to the time-recursive approach used, the initial values $X_c(l,t)$ and $X_s(l,t)$ in the LAIs are reset to zeros every N clock cycles.

The circular shift array in the middle of the system is an $N \times 1$ shift register array as shown in Fig. 9. This special shift register array loads an $N \times 1$ data vector from the LAI every N clock cycles, then it will shift the data circularly and send the data to the $LAII$ every clock cycle. There are three inputs in $LAII$, $\delta_c(k,l,t)$, $X_c(k,l,t)$ and $X_s(k,l,t)$, where the $\delta_c(k,l,t)$ comes from the circular shift array, and $X_c(k,l,t)$ and $X_s(k,l,t)$ from the shift register arrays located behind the $LAII$. We divide the $LAII$ into two groups: the $LAII_{even}$ and $LAII_{odd}$. Each includes $N/2$ lattice modules as shown in Fig. 2. The $LAII_{even}$ contains only those lattice modules for even transformed components k, while $LAII_{odd}$ contains only the odd lattice modules. It should be noticed that this system contains two LAI and only one $LAII$. The shift register array contains $2N \times N$ registers. Their operations are shown in Fig. 10.

The following is to illustrate how this parallel lattice structure works to obtain the 2-D DCT and DSCT of 2-D input successive frames. All the initial values of the circular shift matrix I ($CSMI$), circular shift array, and shift register array are set to zeros. The input data sequence $x(m,n)$ sequentially shifts row by row into the $(N+1) \times N$ $CSMI$. First we calculate the difference between the t'th row and the $(t+N)$'th row data vector of the $CSMI$. The two resulting combinations of the input sequences, $x(t+N,n) - x(t,n)$ and $-x(t+N,n) - x(t,n)$ for $n = 0,1,2,..,N-1$, are used as the input sequences for the lattice array Is, which consist of $2N$ lattice modules to calculate the 1-D DCT for $\{x(t+N,n)-x(t,n)\}$ and $\{-x(t+N,n)-x(t,n)\}$. The upper LAI is for the even transformed components k and the lower one for odd k. Suppose the data of the input vectors arrive serially per clock cycle, it takes N clock cycles to obtain the $\delta_c(k,l,t)$ for both of the input sequences. At the N'th cycle, the N transformed data $\delta_c(k,l,t)$ are loaded into the circular shift arrays, CSA, which will shift circularly and send the data out of the register into the $LAII$ for different k components every clock cycle. In $LAII$, $X_c(k,l,t+1)$ and $X_{sc}(k,l,t+1)$ are evaluated according to (31) and (35). Since $LAII_{even}$ and $LAII_{odd}$ have only $N/2$ modules, every $\delta_c(k,l,t)$ is floating for $N/2$ clock cycles. It is noted that a specific 2-D transform data $X_c(k,l,t+1)$ and $X_{sc}(k,l,t+1)$ are updated recursively every N clock cycles from $X_c(k,l,t)$ and $X_{sc}(k,l,t)$. Therefore the outputs of the $LAII$ are sent into the shift register array (SRA) where data are delayed by N clock cycles. Each SRA contains $N/2$ shift registers each with length N. The data in the rightest registers are sent back as the $X_c(k, l, t)$ and $X_{sc}(k, l, t)$ of $LAII$. At the N^2 clock cycle, the 2-D DCT and DSCT of the 0'th frame are available. After this, the 2-D transformed data of successive frames can be obtained every N clock cycles.

The Post-Matrix Method

The intermediate value $\delta_c(k, l, t)$ in (38) can be rewritten as

$$\delta_c(k, l, t) = (-1)^k \dot{X}_c(t + N, l) - \dot{X}_c(t, l). \tag{39}$$

That is, we can calculate the 1-D DCT of the t'th row and $(t + N)$'th row of the input frame individually, then perform the summation later on. Our approach is to send the input sequence $x(m, n)$ row by row directly into the LAI. It takes N clock cycles for the LAI to complete the 1-D DCT of one row input vector, then the array sends the 1-D DCT data in parallel to the $CSMII$ as shown in Fig. 11. The operations of $CSMII$ are shown in the lower part of Fig. 8. At the output of the $CSMII$, the 1-D transformed data of the $(t + N)$'th row and t'th row are added together according to (39) depending on the sign of the k components (see Fig.11). Then the results are sent to $CSAs$, $LAII$, and $SRAs$, whose operations remain intact as in the pre-matrix method. The whole structure is demonstrated in Fig. 11. Therefore, by transforming the input data first, we can implement the 2-D DCT by using only two 1-D DCT lattice arrays and retain the same pipeline rate. The total numbers of multipliers and adders needed for the post-matrix method are $8N$ and $10N$ respectively.

The Block 2-D DCT Architecture

In most image processing applications, the 2-D DCT are executed block by block instead of in successive frames [39]. We will show how to apply the frame-recursive concept to obtain the block 2-D DCT. It will be easier to understand if we use an example to show how to obtain the 0'th frame 2-D DCT in the post-matrix moving frame 2-D DCT architecture. Recall that the $CSMII$ in Fig. 11 is used to store the transformed data $\dot{X}_c(t, l)$ of the previous input row vectors. Since the 0'th frame is the first input data frame, all the values in the $CSMII$ are set to zeros. Corresponding to (39), this means that the second terms $\dot{X}_c(t, l)$ are zeros. When the $(N - 1)$'th row data vector (the last row of the 0'th frame) arrive, the 2-D DCT of the 0'th input data frame is obtained. During this initial stage, the 2-D DCT of the 0'th frame obtained by the moving frame approach is equal to the block 2-D DCT of the 0'th frame. Therefore, if we want to compute the 2-D DCT of the N'th frame, then all the values in the $CSMII$ are resetting to zeros when the first datum in the N'th data frame (i.e. $x(N, 0)$) arrives. That is, we can obtain the block 2-D DCT by reset the values of the $CSMII$ every N^2 cycles. This means that the $CSMII$ in Fig. 11 is redundant and the second terms $\dot{X}_c(t, l)$ in (39) are zeros. So, the intermediate value of $\delta_c(k, l, t)$ can be rewritten as

$$\delta_c(k, l, t) = (-1)^k \frac{2}{N} C(l) \sum_{n=0}^{N-1} x(t + N, n) \cos \left[\frac{\pi(2n + 1)l}{2N} \right]. \tag{40}$$

The block 2-D DCT architecture is shown in Fig. 12. Corresponding to our frame-recursive algorithm, we obtain another block of input data every

N^2 clock cycles. Note that this is also the total time required for all the N^2 data to arrive in a transmission system.

The following is an example to calculate block 2-D DCT for the 0'th frame. When row data vectors arrive, the LAI performs the 1-D DCT on them. Every N clock cycles, after the last datum of each row $x(m, N - 1)$ is available, the LAI completes the 1-D DCT for every row and sends the N 1-D DCT transformed data to the two length-N CSAs. The upper CSA translates the intermediate value $\delta_c(k, l, t)$ to the lattice array II_{even}, as do the lower CSA except that the signs of the output of the CSA are changed before being sent to the lattice array II_{odd}. The operations of the lattice array II and the SRA are the same as those in the previous methods. As we can see, the hardware complexity of our block 2-D DCT architecture is as simple as the row-column methods. Moreover, our system can be operated in a fully pipelined manner.

Comparisons

In the following, the comparisons between our 2-D DCT block structure and those of others are based on the number of multipliers, adders and speed. For the sake of clarity, we divide the algorithms into two groups: parallel input parallel output ($PIPO$) and serial input parallel output ($SIPO$). The fast algorithms presented by Vetterli and Nussbaumer [5, 37], Duhamel and Guillemot [38], and Cho and Lee [23] belong to the former class. Vetterli's algorithm [37] mapped the 2-D DCT into a 2-D cosine DFT and sine DFT through a number of rotations, and the 2-D DFT are computed by Polynomial Transform (PT) methods [37]. Vetterli's method can reduce the number of multipliers more than 50% in comparison to the row-column method based on Chen's algorithms [7] and has a comparable amount of additions . Duhamel *et al* [38] present a PT-based algorithm which uses the direct DCT approach. This direct PT 2-D DCT method provides a 10% improvement in both the numbers of additions and multiplications compared to Vetterli's result [37] but it requires complex computations. Cho and Lees' algorithm is a direct 2-D method which employs the properties of trigonometric functions. The number of multipliers are the same as that of Duhamel's, but the structure is more regular, and only real arithmetic is involved. Up to now, the best results for the first $PIPO$ systems in terms of the number of multipliers are $(N^2 + 2N + 2)$, which were obtained by Duhamel and Guillemot, as well as by Cho and Lee. But assuming that all the N^2 input data arrive at the same time is not practical in communication systems. The data waiting time is N^2 which is always neglected in these approaches.

The systolic array approaches proposed by Lee and Yasuda [8], Ma [11], and Liu-Chiu belong to the $SIPO$ method. Lee-Yasuda presented a 2-D systolic DCT/DST array algorithm based on an IDFT version of the Goertzel algorithm via Horner's rule in [8]. The latest systolic array algorithm for 2-D DCT was proposed by Ma [11], where he showed two systolic architectures of 1-D DCT arrays based on the indirect approach proposed by Vetterli-Nussbaumer [37], then he exploited the 2-D DCT systolic array by using the features of the two 1-D DCT systolic arrays. This method requires $(N + 1)$ 1-D DCT structures and the total number of time steps is $(N^2 + 2N + 2)$ [11]. We call the block 2-D DCT structure shown in Fig. 12, based on the Liu-Chiu2 module [10], Liu-Chiu2D. This needs only two 1-D DCT, and the total time steps are N^2. The comparisons regarding their inherent characteristics are given in Table 4. In general, the $SIPO$ method is more workable in hardware implementations. Our structure requires fewer

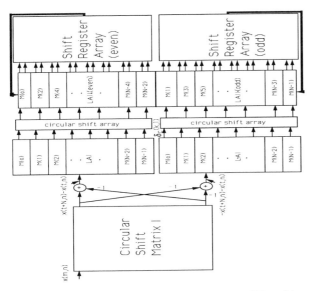

Figure 7: The pre-matrix moving frame 2-D DCT architecture.

	R-C method on Chen	Duhamel et al.	Cho-Lee	Ma	Ours
No. of multipliers	$2N^2\ln(N)$ $-6N^2/2 + 8N$	N^2 $+2N+2$	N^2 $+2N+2$	$4N$ $*N+1$	$8N$
Throughput	$N+$ transposition	$2N$	N	$2N+1$	N
Limitation on N	power of 2	power of 2	power of 2	no	no
Commun.	global	global	global	local	local
I/O	$PIPO$	$PIPO$	$PIPO$	$SIPO$	$SIPO$
Approach	indirect	direct	direct	indirect	direct

Table 4: Comparisons of different 2-D DCT algorithms.

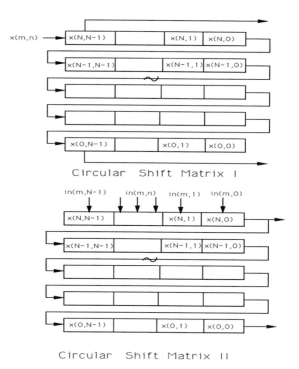

Figure 8: The circular shift matrix (CSM) I and II.

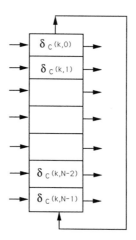

Figure 9: The circular shift array (CSA).

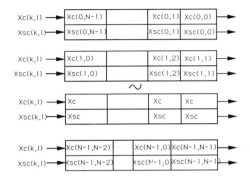

Figure 10: The Shift Register Array.

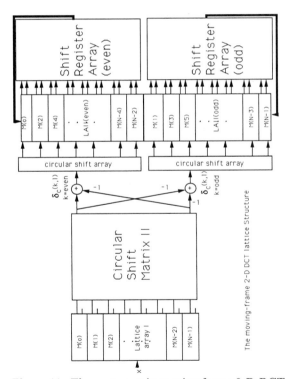

Figure 11: The post-matrix moving frame 2-D DCT architecture.

156

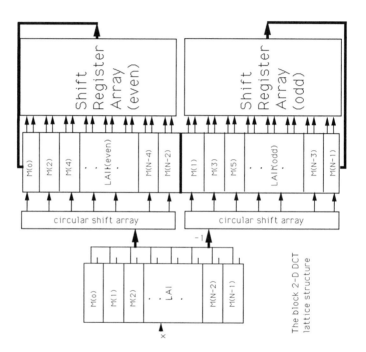

Figure 12: The block 2-D DCT architecture.

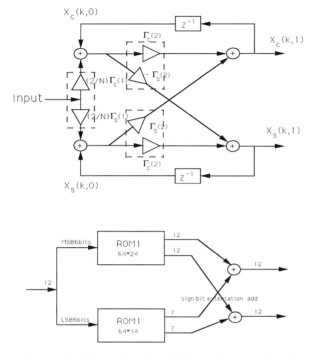

Figure 13: The realization of the lattice module using three ROMs.

multipliers than Ma's structure and is highly regular, systematic, and uses only local communications. In addition, this lattice 2-D DCT architecture can be generated from the 1-D DCT lattice array without modifications.

VLSI IMPLEMENTATOION OF THE DCT/DST LATTICE STRUCTURES

The VLSI implementation of the lattice module using the distributed arithmetic approach is described in this section. The chip was designed for real-time processing of 14.5-MHz sampled video data. Fabricated by using $2\mu m$ double-metal CMOS technology, the lattice module contains approximately 18,000 transistors, which occupy a $5.6 \times 3.4mm^2$ area. The 40-pad die size is $6.8 \times 4.6mm^2$. It has been tested to be fully functional.

Distributed-Arithmetic based Implementation

We will focus our discussion on the 8 point DCT with 8-bit input signals and 12-bit output signals using two's complement binary number system. We implement the lattice module using ROMs as shown in Fig. 13. Each dashed box is realized using a ROM with one input and two outputs. Fig. 13 illustrates the realization of each ROM when the number of bits of the input signal is 6 bits. Using this method, the ROM size of each lattice array is 13824 bits and the number of adders needed is 10. When the number of bits of the input signal is reduced, the ROM size is reduced but the number of adders is increased. We implement our system based on the schematic

158

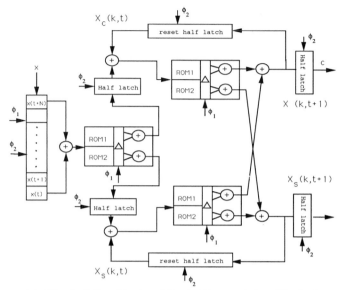

Figure 14: The building blocks of the lattice module with clocks.

diagram for each lattice module as shown in Fig. 13. The adder is a 12-bit carry lookahead full adder/subtracter which is constructed using three 4-bit carry-look-ahead adders. Since, ROMs need less area than general purpose multipliers and can achieve a higher speed, circuit implementations using this approach can be used for very high speed video signal processing.

Design of the Building Blocks

The main building blocks of the lattice structure are ROMs, adders, shift registers, and latches. The ROM design will be described below.

ROM Implementation

Most existing ROMs are implemented based on the precharge concept, that is, the bit lines are precharged high during the precharge phase, and then the selected word lines discharge some of the bit lines according the coefficients stored during the evaluate phase. In order to reduce the ROM access time, we use a novel ROM design [32]. Fig. 15 shows the detail of each cell in the ROM.

A simple inverter with a feedback transistor and a transmission gate controlled by phase ϕ_1 is used as a sense amplifier at the output of the bit-lines. We precharge the bit lines to an intermediate voltage between GND and Vdd, and use n-channel transistors to either charge the bit line from this intermediate voltage to Vdd-Vt or discharge it to GND, during the evaluate phase. In this scheme, the array is fully populated; . *i.e.* the number of n-channel transistors in the array is MN, where M is the number of word lines and N is the number of bit lines. A 'zero' is stored at a particular

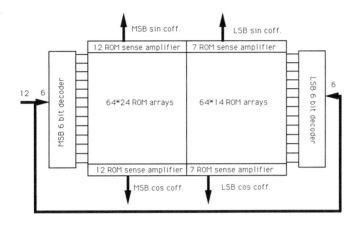

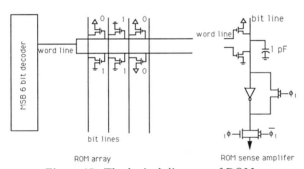

Figure 15: The logical diagram of ROM.

location, by connecting the n-channel transistor at that location to Vdd; a 'one' is stored by connecting the transistor to GND. The cell size is only $13\lambda \times 16\lambda$.

In our distributed-arithmetic scheme, the multiplication of the 12-bit input number with a 12-bit sine or cosine coefficient is performed by two ROMs each with 6-bit inputs and two adders. This reduces the chip area needed to implement multiplication with fixed coefficients. The ROM includes two 6-bit decoders and two small ROMs as shown in Fig. 14. The 12-bit input is divided into two parts; the most significant 6-bits of the input are used to generate the coefficients for small ROM1 and the least significant 6-bits are used for small ROM2. The final result of the multiplication is obtained by adding the outputs of small ROM1 with a shifted version of the outputs of small ROM2. We only store the most significant 7-bit result of the multiplication at ROM2. The sizes of small ROM1 and ROM2 are 64 words by 24 bits and 64 words by 14 bits respectively.

In order to improve the ROM access time, each 6-bit decoder is implemented as a tree consisting of two 3-bit decoders and linear array of 64 AND gates. The delay time for this 6-bit decoder is 8.55ns, while a straightforward implementation would have a delay of 20.40ns. The outputs of the 64 AND gates form the word lines of the ROM array The physical size of the final ROM is $1292\lambda \times 1646\lambda$ which is much smaller than the area needed by a general purpose multiplier. The total ROM access time is 20ns.

Chip Realization and Simulations

Having realized the symbolic layout of the individual blocks, the next issue is to integrate all these components efficiently. This includes three ROMs, eleven adders, four half-latches, two reset controlled half-latches and one shift register. ROM2 and ROM3 are rotated by 90 and 270 degree respectively to simplify inter-component routing. This chip accepts 8-bit input signals and produces 12-bit DCT and DST coefficients every 100ns. The physical layout of the lattice module chip is depicted in Fig. 16. There are 18000 transistors in the chip, most of which are used in the three fully-populated ROMs. The total size of the active circuitry is $5400\lambda \times 3400\lambda$. This is fabricated in a die of size $6800\lambda \times 4600\lambda$ and packaged in a 40-pin chip.

CONCLUSIONS

In this Chapter, we have studied the problem of developing efficient VLSI architectures for real-time computation of the discrete sinusoidal transforms, especially the DCT, DST, DHT, DFT, LOT, CLT.

We proposed a "time-recursive" approach to perform those transforms that merges the buffering and transform operations into a single unit. The transformed data are updated according to a recursive formula, whenever a new datum arrives. Therefore the waiting time required for other algorithms is completely eliminated. The unified lattice structure for time-recursive transforms is proposed. The resulting architectures are regular, modular, locally-connected and better suited for VLSI implementations. There is no limitation on the transform size N and the number of multipliers required is linear function of N. The throughput of this scheme is one input sample per clock cycle.

We also look at the problem of applying the time-recursive approach to

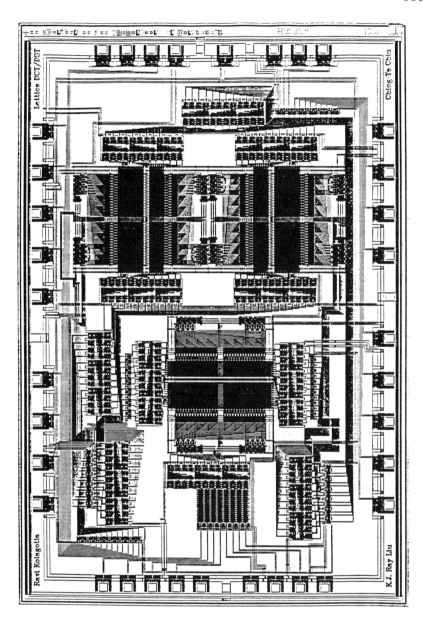

Figure 16: The physical layout diagram of the lattice module.

the two-dimensional DCT. Our fully-pipelined 2-D DCT lattice structure requires only two 1-D DCT arrays which are extended directly from the 1-D DCT structure without transposition. The system throughput rate is N clock cycles for a $N \times N$ successive input data frame.

The VLSI implementation of the lattice module based on the distributed arithmetics is also described. The chip has been fabricated under $2\mu m$ double-metal CMOS technology and tested to be fully functional with a data processing rate of 116Mb/s.

In conclusions, we propose a new point of view in the VLSI implementation of the discrete sinusoidal transforms. Instead of using different implementation techniques to improve the realization of FFT-like algorithms, we start from the serially available property of signals in the transmission and scanning systems to improve the operation speed of the systems. All these discrete sinusoidal transforms can be generated by using same lattice architectures with different multiplication coefficients. The fundamental dual generation relations between the DCT/DST, DHT/DFT and LOT/CLT can also be derived. We have shown that this approach is a very promising method for the applications of video communications.

References

[1] N. Ahmed, T. Natarajan, and K. R. Rao, "Discrete cosine transform," IEEE Trans. Comput., vol. C-23, pp. 90-93, Jan. 1974.

[2] R. N. Bracewell, "Discrete Hartley transform," J. opt. Soc. Amer., vol. 73, pp. 1832-1835, Dec. 1983.

[3] R. Yip and K. R. Rao, "On the computation and effectiveness of discrete sine transform," Comput. Elec. Eng., vol. 7, pp. 45-55, 1980.

[4] H. S. Hou, "A fast recursive algorithm for computing the discrete cosine transform," IEEE Trans. Acoust., Speech, Signal Processing, vol. ASSP-35, pp 1455-1461, Oct. 1987.

[5] M. Vetterli and H. Nussbaumer, "Simple FFT and DCT algorithm with reduced number of operations," Signal Processing, vol. 6, no. 4, pp. 267-278, Aug. 1984.

[6] M. Vetterli, and A. Ligtenberg, "A discrete Fourier-Cosine transform chip," IEEE Journal on Selected Areas in Communications, vol. SAC-4, No. 1, pp. 49-61. Jan. 1986.

[7] W. H. Chen, C. H. Smith, and S. C. Fralick, ' 'A fast computational algorithm for the discrete cosine transform," IEEE Trans. Communication, vol. COM-25, pp. 1004-1009, Sept. 1977.

[8] B. G. Lee, "A new algorithm to compute the discrete cosine transform," IEEE Trans. Acoust., Speech, Signal Processing, vol. ASSP-32, pp 1243-1245, Dec. 1984.

[9] N. H. Lee and Y. Yasuda, "New 2-D systolic array algorithm for DCT/DST," Electron. Lett., 1989, 25, pp. 1702-1704.

[10] K. J. R. Liu, and C. T. Chiu, "Unified Parallel Lattice Structures for Time-Recursive Discrete Cosine/Sine/Hartley Transforms," IEEE Trans. on Signal processing, vol. 41, No. 3, pp. 1357-1377, March 1993.

[11] W. Ma, "2-D DCT systolic array implementation," Electronics Letters, Vol. 27, No. 3, pp. 201-202, 31st Jan. 1991.

[12] H. J. Nussbaumer, and P. Quandale, "Fast Polynomial Transform Computation of 2-d DCT," Proc. Int. Conf. on Digital signal Processing, Florence, Italy, pp. 276-283, 1981.

[13] N. I. Cho and S. U. Lee, "Fast algorithm and implementation of 2-D Discrete Cosine Transform," IEEE Trans. on Circuits and Systems, vol. 38, No. 3, pp. 297-305, March. 1991.

[14] K. Rose, A. Heiman, and I. Dinstein, "DCT/DST Alternate-Transform Image Coding," IEEE Trans. on Communication, vol.38, No. 1, pp. 94-101, Jan. 1990.

[15] M. T. Sun, T. C. Chen, and A. M. Gottlieb, "VLSI implementation of a 16×16 Discrete Cosine Transform," IEEE Trans. on Circuits and Systems, vol. 36, No. 4, pp. 610-617, Apr. 1989.

[16] S. A. White, "High-Speed Distributed-Arithmetic Realization of a Second-Order Normal-Form Digital Filter," IEEE Trans. on Circuits and Systems, vol. 33, No. 10, pp. 1036-1038, Oct. 1986.

[17] S. A. White, "Applications of Distributed-Arithmetic to Digital Signal Processing: A Tutorial Review," IEEE ASSP Magazine, pp. 4-19, July. 1989.

[18] C. Chakrabarti, and J. J'aJ'a, "Systolic architectures for the computation of the discrete Hartley and the discrete cosine transforms based on prime factor decomposition," IEEE Trans. on Computer, vol. 39, No. 11, pp. 1359-1368, Nov. 1990.

[19] Z. Wang, "Fast algorithms for the discrete W transform and for the discrete Fourier transform," IEEE Trans. Acoust., Speech, Signal processing, vol. ASSP-32, Aug. 1984.

[20] R. Yip and K. R. Rao, "On the shift property of DCT's and DST's," IEEE Trans. Acoust., Speech, Signal Processing, vol. ASSP-35, No. 3, pp. 404-406, March. 1987.

[21] H. V. Sorenson, et. al., "On computing the discrete Hartley transform," IEEE Trans. Acoust., Speech, Signal Processing, vol. ASSP-33, No. 4, pp. 1231-1238, Oct. 1985.

[22] S. B. Narayanan and K. M. M. Prabhu, "Fast Hartley transform pruning," IEEE Trans. Acoust., Speech, Signal Processing, vol. ASSP-39, No. 1, pp. 230-233, Jan. 1991.

[23] N. I. Cho and S. U. Lee, "DCT algorithms for VLSI parallel implementations," IEEE Trans. Acoust., Speech, Signal Processing, vol. ASSP-38, No. 1, . 1899-1908, Dec. 1989.

[24] L. W. Chang and M. C. Wu, "A unified systolic array for discrete cosine and sine transforms," IEEE Trans. Acoust., Speech, Signal Processing, vol. ASSP-39, No. 1, pp. 192-194, Jan. 1991.

[25] E. A. Jonckheere and C. Ma, "Split-radix fast Hartley transform in one and two dimensions," IEEE Trans. Acoust., Speech, Signal Processing, vol. ASSP-39, No. 2, pp. 499-503, Feb. 1991.

[26] N. Takagi, T. Asada, and S. Yajima, "Redundant CORDIC Methods with a Constant Scale Factor for Sine and Cosine Computation," IEEE Trans. on Computers, vol. 40, No. 9, pp. 989-995, Sep. 1991.

[27] A. V. Oppenheim and R. W. Schafer, **Discrete-Time Signal Processing**, Prentice Hall, 1989.

[28] S. A. White, "High-Speed Distributed-Arithmetic Realization of a Second-Order Normal-Form Digital Filter," IEEE Trans. on Circuits and Systems, vol. 33, No. 10, pp. 1036-1038, Oct. 1986.

[29] R. Young, and N. Kingsbury, "Motion Estimation using Lapped Transforms," IEEE ICASSP Proc., pp. III 261-264, March, 1992.

[30] R. D. Koilpillai, and P. P. Vaidyanathan, "New Results on Cosine-Modulated FIR filter banks satisfying perfect reconstruction," IEEE ICASSP, 1991, pp. 1793-1796.

[31] C. T. Chiu, and K. J. R. Liu, "Real-Time Parallel and Fully-Pipelined Two-Dimensional DCT Lattice Structures with Application to HDTV Systems," IEEE Trans. on Circuits and Systems for Video Technology, pp. 25-32, March 1992.

[32] L. A. Glasser, and D. W. Dobberpuhl, "The Design and Analysis of VLSI Circuits," Addison Wesley, 1985.

[33] Neil H. E. Weste, and Kamran Eshraghian, "Principles of CMOS VLSI Design, A Systems Perspective," Addison Wesley, 1985.

[34] C. Chakrabarti, and J. JáJá, "Systolic architectures for the computation of the discrete Hartley and the discrete cosine transforms based on prime factor decomposition," IEEE Trans. on Computer, vol. 39, No. 11, pp. 1359-1368, Nov. 1990.

[35] P. M. Cassereau, D. H. Staelin, and G. D. Jager,"Encoding of Images Based on a Lapped Orthogonal Transforms," IEEE Trans. on Communications, Vol. 37, No. 2, pp. 189-193, Feb. 1989.

[36] D. Slawecki and W. Li, "DCT/IDCT Processor Design for High Data Rate Image Codings," IEEE Trans. on Circuits and Systems for Video Technology, pp. 135-146, June 1992.

[37] M. Vetterli, "Fast 2-D discrete Cosine Transform," IEEE ICASSP Proc., pp. 1538-1541, March. 1985.

[38] P. Duhamel and C. Guillemot, "Polynomial Transform computation of the 2-D DCT," IEEE ICASSP Proc., pp. 1515-1518, March. 1990.

[39] S. Cucchi, and F. Molo, "DCT-based Television Codec for DS3 digital Transmission," SMPTE Journal, pp. 640-646, Sep. 1989.

[40] W. Li, "A new algorithm to compute the DCT and its inverse," IEEE Trans. on Signal Processing, vol. 39, n0. 6, pp. 1305-1313, June, 1991.

6

DESIGN AND PROGRAMMING OF

SYSTOLIC ARRAY CELLS FOR SIGNAL PROCESSING

Ross Smith
University of Illinois at Chicago
Department of Electrical Engineering
and Computer Science (M/C 154)
Chicago, IL 60607

Gerald Sobelman
University of Minnesota
Department of Electrical Engineering
Minneapolis, MN 55455

1. INTRODUCTION

Many signal processing and matrix algorithms require large amounts of computation, yet the computations are inherently regular and could easily be done in parallel. In 1978, H. T. Kung and C. E. Leiserson [1] introduced systolic arrays, a parallel architectural approach that could use VLSI technology to exploit these algorithms. The word systolic was chosen because a systolic array is controlled by a global clock, so the data pulses through the array synchronously.

The main ideas behind systolic arrays can be explained using one-dimensional convolution. A three-point convolution has the following equation

$$y_i = w_0 * x_i + w_1 * x_{i-1} + w_2 * x_{i-2}$$

Note the inherent regularity in this problem. Each x_i is multiplied once by each of the w_i's, and each w_i is multiplied by the x_i's in order. A linear systolic array for this algorithm is shown in Figure 1 [2]. Three cells are used, one for each weight, w, and three data streams are used, one each for w, x, and y. Each cell consists of a multiplexer, four registers, and a multiplier/accumulator, or MAC which executes the inner product, ACC = ACC + (X * Y). In this array, the w_i's are sent in a repeated pattern of w_1, w_2, w_3. The x_i's are sent in order from the first point to the last, and upon completion of a convolution, a y_i is ejected and sent out on a data stream. Each cycle, the following events occur in every cell. In the MAC, X and Y are multiplied together and added to ACC. The result is stored in ACC. New values for X and Y are loaded from w_i and x_i, and all the register values are shifted left one register.

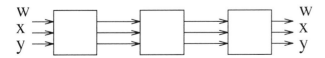

Figure 1. A Linear Array For Convolution (After [1]).

This convolution example illustrates four points about systolic arrays [1]: points 1 & 2 are concerned with exploiting the parallelism, and points 3 & 4 are concerned with exploiting VLSI technology.

(1) A systolic array makes multiple use of each input data item. Because of this property, systolic arrays can achieve high throughput with modest I/O bandwidth. In the convolution example, no matter how many cells are used in the linear array, the I/O bandwidth requirements are the same for the host.

(2) A systolic array uses extensive concurrency. The processing power of a systolic architecture comes from concurrent use of many simple cells rather than sequential use of a few powerful processors. In the convolution example, all resources were used every cycle.

(3) There are only a few types of simple cells. Thus, the designer's task is eased, because the cells designed are replicated many times. The convolution example used only one type of cell.

(4) The data and control flow is simple and regular. Pure systolic systems do not use long-distance or irregular wires for data communication. This enables the design to be modular and expandable. The convolution example had very regular data flow.

Since systolic arrays were first introduced, many architectures have been proposed. A variety of topologies have been used: linear, rectangular, tree, triangular, or hexagonal. The arrays are either dedicated to a specific application or programmable; programmability enables an array to execute a family of algorithms. Most of the systolic arrays implemented have been programmable, and have used one of three approaches. One approach uses very powerful cells with high-speed communication and arithmetic. They have a few cells, but hundreds of chips per cell [3, 4]. For example, the Warp processor from CMU has a 32-bit floating-point multiplier and a 32-bit floating-point ALU, but originally used 252 chips per cell [5, 6]. The other extreme uses cells which operate on a single bit. With this type of cell, many cells are used per array, and tens or even hundreds of cells can fit on a chip. An example of this is the Distributed Array Processor [7]. The approach in the middle uses cells which have bit-parallel arithmetic, but only a few chips per cell. Thus, the cells are much more powerful than bit-oriented cells, but less expensive than the most powerful cells. The first implementation of an inexpensive systolic array cell of this type was the Programmable Systolic Chip [8].

Although many systolic array architectures and algorithms for digital signal processing have been proposed, only a few experimental systolic arrays have been implemented because of the high development costs associated with their design and programming. Essentially, the design and programming of systolic arrays is limited by a lack of design and programming building blocks, tools, and techniques. Designing a systolic array is expensive because the hardware designer uses several techniques that increase throughput, but increase also design complexity: partitioning, parallelism, pipelining within the cell, pipelining of the cells, and synchronous communication. These hardware techniques also make programming a systolic array very complex. The complexity of designing an array could be reduced if building block cells were available for the designer who wants to use a systolic array, but does not want to design a cell. The complexity of designing a cell could be reduced if a design approach and tools were available for designing and simulating the cell and array. The programming complexity could be reduced if a programming approach and software tools for developing programs were available.

Each section in this chapter examines one of these areas. Section two presents a building block cell that was designed and implemented. Issues involved in using this cell as a building block in an array are also examined. Section three examines how to program this cell. Section four examines how to design systolic array cells. Finally, section five describes a second cell, which has several unique features for systolic arrays.

2. THE SYSTOLIC ARRAY CONTROLLER CHIP: SAC

2.1 Introduction

It is much more convenient to use a systolic array that can be used for a wide variety of signal processing problems rather than to build a dedicated array that can only be used for a single systolic algorithm. A systolic array cell may be made flexible by making it programmable. Because of their different applications, programmable cells differ in terms of arithmetic, routing, memory and control capabilities. For example, complex operations such as division and square root functions may only be required for the cells along the boundaries of the array. Cells for applications requiring only a limited dynamic range could use fixed-point arithmetic, while those requiring a large dynamic range would have floating-point capabilities.

From the point of view of an array designer, it would be useful to have a set of "building block" cells, that is, a set of standard, components for assembling systolic array systems, similar to the way the microprocessor-based system design components are presently used. A cell-based building block approach to systolic array implementation has been investigated by [8-11]. The cells in these papers have specially designed arithmetic hardware, whereas in this chapter we use a custom controller chip and an existing mathematics processor chip to form a low-cost, high-performance two-chip cell. Our particular cell-based building block approach also differs from that of [12], who designed a chip which could be used as a building block for constructing the individual cells themselves.

For applications of practical size a large number of cells are needed. However, in order to do this, the individual cells themselves should be relatively simple so as to keep the overall system cost low. However, two forces work against the desire for a simple cell. First, the arithmetic capabilities of each cell must be powerful enough to meet the application's needs. Second, a cell intended for use in a variety of applications requires a flexible, and hence more complex, cell. Powerful and flexible cells composed of a large number of chips lead to a high overall cost.

In the two-chip per cell approach (Figure 2), a commercially-available mathematics processor chip (the MAC) provides the arithmetic capabilities, while a specially designed controller chip (the SAC) provides the required control sequencing and data routing functions. The cost is kept relatively low because only two chips

168

per cell are used, and yet the cell has considerable hardware resources and programming flexibility. Moreover, cell design time is reduced by not having to re-design the computational functions that are already present in the high-performance commercial mathematics processors that are now widely available.

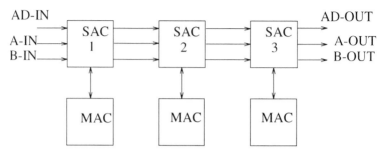

Figure 2. A Linear SAC/MAC Array.

2.2. Cell Architecture

A signal processing cell was developed that uses the 16-bit, fixed-point arithmetic, single-port NCR45CM16 CMOS Multiplier/Accumulator (MAC) [13], together with a custom-designed router/controller chip, which we call the Systolic Array Controller (SAC). We wanted the SAC/MAC cell was to be an inexpensive, yet flexible and efficient cell. A large portion of the cost of a chip is determined by the total design time required. By using a commercially available MAC (multiplier/accumulator), the total cell design time was reduced considerably. Moreover, by using a standard cell design methodology, IBM's MVISA system [14], the design time for the SAC chip was minimized. The MVISA system has a pre-defined library of macro cells together with all of the required CAD tools for chip implementation. In addition, it also incorporates the well-established LSSD-based testing methodology [15]. The SAC chip has been fabricated by IBM. It is implemented in a 2-micron NMOS technology, has a die area of 4.2 mm by 4.2 mm, and has 72 signal I/O pins.

Now, consider the other two design goals, namely flexibility and efficiency. Flexibility may be obtained through cell programmability as well as by accommodating a large set of interconnection configurations. Efficiency, on the other hand, requires that the hardware resources are well matched to the required computational tasks. The SAC design (Figure 3) together with the MAC includes all the features systolic arrays need [8]: bit-parallel multiply/accumulate function, local RAM, high-bandwidth communication, and microprogrammed control. The importance of each feature is explained below, together with a description of its implementation in the SAC/MAC cell.

bit-parallel multiply/accumulate function - The multiply/accumulate function is a central feature of most signal processing algorithms. We used the NCR45CM16 chip, which has a 40-bit accumulator and two 16-bit input registers. It can multiply/add, multiply/subtract, and clear. A 16 by 16 bit fixed-point signed two's complement multiplication and 40-bit accumulation can be done in a single 220 ns cycle, while one iteration of a pipelined series of inner products can be done in two cycles.

on-chip RAM - The placement of a RAM on the SAC chip allows the cell to be used for a wide variety of systolic algorithms. This RAM holds programs and data. Storing programs in the RAM eliminates the need to pass instructions to a cell during program execution. In addition, storing data in the RAM will reduce cell I/O requirements if an operand needed is in the RAM. A 64 by 18 bit static RAM is used for storing both control and data. (An instruction word has 18 bits and a data word has 16 bits). An important design optimization was possible due to the fact that the MAC has a relatively long cycle time (220 ns) compared to that of the internal RAM (70 ns). Therefore, the RAM may be used twice during each cycle, once to read a control word and once to read or write a data word.

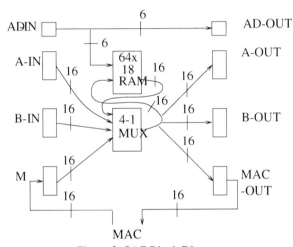

Figure 3. SAC Block Diagram.

high-bandwidth communication - In the case of a two-chip cell implementation, high-bandwidth communication is composed of two components, inter-cell communication and intra-cell communication. First, consider the case of inter-cell communication. In general, wide data paths allow an efficient and continuous flow of data between the cells. However, pin limitations preclude the use of a large number of wide I/O ports. The SAC has two 8-bit data input ports, two 8-bit data output ports, one 3-bit address input port, and one 3-bit address output port. We were able to place six SAC/SAC ports onto the 72-pin SAC chip through the use of half-word ports. Each port is used twice each cycle, once to load the least significant bits and once to load the most significant bits. Obviously, the use of half-word ports reduces the number of I/O pins required. However, the important point is that a straightforward program control over the I/O operations is maintained because a full word is transferred during each cycle. Now consider the case of intra-cell communication. The SAC/MAC cell contains one 16-bit bidirectional port for routing data between the SAC and the MAC as well as one 16-bit 4-1 multiplexer for routing data internally within the SAC. As was the case for the I/O ports, the multiplexer is also used twice each cycle, once for writing A-IN or M to RAM and/or A-OUT, and once for writing A-IN, B-IN or RAM to B-OUT and/or the MAC.

microprogrammed control - Microprogrammed control provides a great deal of flexibility. Moreover, cell performance is enhanced by allowing the different modules of the chip to operate in parallel under program control. In the case of the SAC, during every cycle the MAC, RAM, program counter, and all of the registers may be operated concurrently.

In summary, the SAC/MAC cell includes all the important hardware features outlined above. This was accomplished through a careful analysis of the design trade-offs that could be made in order to achieve the desired objectives and yet keep within the limitations of the available technology. For example, the use of several resources twice during each cycle was an important design feature. In addition, the careful balancing of the computation times for the MAC as opposed to those internal to the SAC lead to an efficient implementation. Finally, some restrictions were imposed of the set of possible routing and control options.

2.3 Performance Benchmarks

Performance comparisons are made difficult by differences in architectural and technological factors. The speed of a DSP or mathematics processor chip is largely a function of the technology used, which in turn depends to a great extent on the year that the chip was manufactured. Thus, it is more useful to look at the number of computations performed per cycle rather than the number of computations per unit time. Comparing architectures from this perspective, the key question becomes, how efficiently is any given mathematics processor chip utilized? The SAC controls the MAC in very efficiently. In fact, aside from the cycles devoted to the initialization process, the SAC/MAC interface is used on every cycle for many programs.

Table 1 illustrates the number of cycles required to execute a representative set of algorithms on various array configurations. Matrix-matrix multiplication is examined using three different systolic arrays, a two-dimensional array, a one-dimensional array and a single SAC/MAC cell.

Algorithm	Array Size	Cycles
N by N multiplication	N by N	$3N+3$
N by N multiplication	N by 1	$3N_2+4N+2$
N by N multiplication	1 by 1	$2N_3+N_2+4$
1 butterfly	1 by 1	11
16-point FFT	4 by 1	182
N point 1-D convolution	N by 1	$3N+5$

Table 1. Execution Times for Various Algorithms.

It should be noted that a systolic array has a low I/O bandwidth for a given computation rate because the array makes multiple use of each data item. For example, the interface to a linear array of SAC/MAC cells runs at 8 MHz, yet a 10-cell array would have a peak throughput of 40 million integer operations per second. In addition, the host/array interface requires only 40 pins, not counting the clock pins. The SAC's performance could be further improved if it contained a larger RAM and had the ability to load two words of data into the RAM each cycle during the initializing stage.

2.4 Conclusions

The two-chip cell approach is a very attractive option for designing cells for systolic arrays. Powerful and flexible systems can be built using this approach without having to re-design readily available mathematics processing hardware and without using a large number of chips per cell. This implementation approach provides advantages to both array and cell designers. For array designers, the SAC/MAC cell is cost-effective, flexible, and has a high throughput. Using only two chips per cell, one can implement one-dimensional and two-dimensional arrays for either application-specific or general-purpose machines. For cell designers, the SAC/MAC approach offers similar benefits. The designer first chooses a MAC to meet the computational requirements of the application, and then he/she can focus solely on communication issues. An added advantage is that an existing SAC design can be modified for use with other MACs. For example, in section 5, another design that uses similar architectural features, but which is designed to work with the AMD 29325 single-chip floating-point processor [16] will be described.

3. TOOLS FOR PROGRAMMING

3.1. Introduction

As with any high-performance multi-processor system, programming a SAC/MAC array is somewhat complex due to the abundance of hardware features that are used to enhance its performance, including on-chip parallelism, pipelining within the cells, pipelining of the cells, and synchronous communication. Previous work has been done on techniques for improving the programming environment for systolic arrays. [17] discussed block processing, a procedure for breaking up algorithms into "a set of 'primitives' which operate on 32 x 32 block of data". [18] examined how pipelining can be handled so that "once a 'generic' systolic algorithm is designed, other versions of the algorithm (for execution on arrays with failed cells, or for implementation using different pipelined processing units) can be systematically derived." Snyder developed Poker [19], a parallel programming environment which can be used as a simulator for developing systolic array hardware and software. [20] demonstrated "how Warp can be programmed as a sequential machine without compromising the efficiency." More recently, [21] developed a high-level language for writing Warp programs, and they also developed an optimizing compiler for the language. See also [22-25] for details on programming systolic arrays.

Below, we present an approach for developing a wide range of systolic array algorithms. Although the discussion is geared specifically towards the SAC/MAC cell, the ideas would also be applicable to many other types of systolic array cell program development situations.

3.2. The Library of Macro Routines

The macros to be described in this section will be presented in a sequence that would reflect their order of usage in typical application programs. First, those macros dealing with the operation of the MAC will be described. Second, those dealing with RAM usage will be defined. Next, the code for the input and output procedures will be determined based on the MAC and RAM usage. Finally, we will show how the macros can be combined.

3.2.1. MAC Programming

A set of three different macros have been defined for MAC usage. These are: vector product, accumulate and read result.

(a) Vector Product

The macro routine for an n-element vector product,

$$ACC = x_0 * y_0 + x_1 * y_1 + \cdots + x_{n-1} * y_{n-1},$$

is composed of the following sequence of microcode instructions: (Note that in this and subsequent program listings, all operations associated with the same line number will be executed concurrently. X and Y are registers in the MAC.

1. $ACC = 0, X = x_0$
2. $Y = y_i$
3. $ACC += X * Y, X = x_i$, *loop until* $AD \equiv 1$
4. $M = ACC[i-j]$

The instructions always increment the program counter, except for step 3, which decrements it unless AD-IN $\equiv 1$. Step 1 clears the accumulator and loads X, and step 2 loads Y. In step 3, X is loaded, and the multiplication/accumulation operation is complete. Steps 2 and 3 form a two-instruction loop that will be repeated n times. Thus, on every cycle, either X or Y will be loaded with a new operand and on every other cycle the new operands will be accumulated. On the nth execution of step 3, X will be loaded with data that will not be used for the current vector product. However, it can be used as an operand for the next vector product. Finally, the MAC will be read in step 4. (Note that the particular group of bits to read, as denoted by the expression $[i-j]$, will be explained later, as a part of the presentation of the MAC read macro.)

(b) Accumulate

An accumulate operation, $ACC += x_0$, requires three instructions:

1. $X = 1$
2. $Y = x_0$
3. $ACC += X * Y$

Hence, simply adding a number, without any multiplication, would require as many cycles as a multiply/accumulate because the number must first be multiplied by 1 and then added. However, a series of additions would be faster, as n additions would require only $n + 2$ cycles.

(c) Read Result

Once a result resides in ACC, it must ultimately be transferred out of the MAC. Because only one 16-bit MAC I/O port is available, more than one individual read operation may be required to extract the entire result. The result after one multiply/accumulate operation using two 16-bit operands requires up to 32 bits for its exact representation. In this case, two words must be read in order to obtain the complete result. For example, if five multiply/accumulates are executed, it is possible that 34 bits would be needed to contain the exact, full precision representation of the result. In that case, three individual read operations would have to be executed in order to obtain the exact representation of the computed result.

3.2.2. Initialization of RAM Contents

Once the MAC control has been determined, the next step is to determine how each SAC's internal RAM will be used. In particular, the RAM must be loaded with data. The basic algorithm used is

> 1. *hold until AD != 1*
> 2. *load until AD ≡ 1, RAM[AD-IN] = A-IN*
> 3. *pass until AD ≡ 1,*
> 4. *loop until AD ≡ 1,*

Although it is not shown in the algorithm, the first two instructions read AD and A, waiting for tokens which indicate when to stop holding, and when to stop loading. The last two instructions pass AD and A data to the next cell in addition to reading the data streams. The time required to load W words into each cell of an N-cell linear array of SACs is $(W + 2)N$.

3.2.3. Data Streams

The MAC and RAM programs determine the operands that are needed. The data streams, on the other hand, specify the manner in which those operands are supplied to the array and how results are to be unloaded from the array. In general, there are two possible options that could be employed. In the first option, the RAM and the B data stream supply the operands and the A stream unloads results. In the second option, both of the data streams are used to supply operands.

In order to set up a data stream, it is necessary to specify certain delay values. Let *reg-in* = A-IN, B-IN, or AD-IN, and *reg-out* = A-OUT, B-OUT, or AD-OUT. Delays d_{in}, d_{out}, and d_{data} are used to measure how long the data resides in *reg-in* (d_{in}), how long data resides in *reg-out* (d_{out}), as well as the delay between consecutive data values (d_{data}). (Note that these delay values are assumed to be in units of integer multiples of the basic cycle time, 220 ns.) The delay in the cell is $d_{cell} = d_{in} + d_{out}$. The first word can be loaded into a cell and transferred out with the following register transfers,

$$i \qquad read\ reg\text{-}in$$
$$i + d_{in} \qquad reg\text{-}out = reg\text{-}in$$

where i and $i + d_{in}$ refer to the instruction number in which the register transfers take place. If the register transfer instructions are used in a loop, then d_{data} must be greater than $MAX(d_{in}, d_{out})$ so as to avoid overwriting data before it has been read. If $d_{data} \geq d_{cell}$ an additional stream can be accommodated. This will not be discussed here, but an example of it is contained in the two-dimensional matrix multiplication program to be presented later. For two-instruction loops d_{data} should equal 2, so that on every other cycle a new word of data is available.

The cells can be synchronized to have a delay of d_{cell} with

$$
\begin{array}{ll}
i & \text{hold until AD} != 1, \text{read AD-IN} \\
i + d_{cell} - 1 & \text{AD-OUT} = \text{AD-IN}
\end{array}
$$

The first instruction makes the cells hold until a 000001 is received on AD-IN. The second instruction controls the delay between cells by controlling how long the 000001 stays in a cell before the next cell can access it.

3.2.4. Program Construction

The macros described above may be combined to form a complete systolic array program. The individual macros would ordinarily be placed in the following sequence:

1. load RAM (optional)
2. synchronize the cells/initialize the MAC/initialize the data streams
3. compute results/send data onto streams
4. unload results

Once the macros are combined the resulting program should be closely examined to see how adjacent iterations interact with one another. Upon careful analysis of the first and last steps of a program, one or two instructions can often be combined in order to save time and memory. A program can be pipelined if the results of the current iteration can be unloaded while the next iteration begins.

3.3. Conclusion

A systematic approach to the development of systolic array programs using the SAC/MAC cell has been described in this section. The approach is centered around the development of a library of macros which implement commonly used program fragments. A particular systolic algorithm of interest can be easily programmed on an array through the appropriate combination of a subset of these macros. In most cases, the programs can be obtained with little or no new microcode required. Using such a macro-based approach, the primary software development tasks to be performed are: programming the MAC, loading the RAM, synchronizing the cells, delaying data, unloading results, combining the macros into the program, and, finally, pipelining the program. For additional details see [26].

4. SIMULATION-BASED DESIGN

4.1. Introduction

Systolic arrays can provide an efficient implementation of signal processing algorithms based on matrix computations [27]. Two programmable systolic array cells for signal processing have been developed by the authors. The initial difficulties that were encountered in designing the first cell led to the development of tools to better manage the design process. These tools allowed us to develop and verify the hardware and algorithms concurrently. For example, given a preliminary architecture, its performance could be examined using different algorithms. Conversely, a single algorithm could be examined using variations of the cell's architecture. Because the algorithms and the architecture are designed together, in the beginning neither the algorithm nor the architecture can be assumed to be correct.

Several excellent design methodologies have been proposed for systolic arrays [25, 28, 29]. Programming tools for systolic arrays also exist [19]. In addition, many special-purpose architectures have been proposed for systolic arrays. In most of these approaches to systolic array design, however, the level of abstraction is quite high; translation of the architecture into actual hardware is not normally specified. On the other hand, design tools exist for chip-level modeling and simulation. However, these tools do not offer insight into the high-level array effects. Tools are needed for the high-level behavior design as well as for the logic and physical design. We have combined these two requirements into a single design approach.

In this section, the approach developed will be described using the Floating-point Processor Controller chip, or FPC. First, the algorithms that were targeted and the constraints and requirements these algorithms imposed will be examined. From these specifications, a preliminary architecture can be obtained. Next, the hardware constraints and their effect on the implementation of the architecture will be described. Finally, the simulator is used to optimize the architecture and the algorithms. For additional references and details see [30].

4.2. Preliminary Architecture

We wanted the FPC to be able to efficiently execute a variety of algorithms. Thus, for the initial specification, cell architectures were examined which could implement convolution, matrix multiplication, matrix triangularization with neighbor pivoting, orthogonal triangularization, and the FFT. By examining these special-purpose designs, a preliminary general-purpose design was specified. Our focus was on the common architectural features of the internal cells designed. These common features suggested the basic form that a general-purpose cell should take.

Preliminary hardware specifications were developed from the hardware required by these algorithms. The minimum design goals were to have floating-point multiplication and addition, two data streams with variable delays, on-chip program and data RAM, and two systolic control streams.

The next step was to choose the hardware that would be used to implement the design. Like the SAC/MAC design, a math chip and a controller chip were used. To eliminate the design time required to develop a math chip, the Advanced Micro

Devices Am29325 32-bit floating-point processor (FPP) was used [16]. The Am29325 can subtract, add, and multiply. The Floating-point Processor Controller (FPC) chip was designed using IBM's MVISA master-image design tools [14]. As in the SAC/MAC design, the FPC controls and routes data to the Am29325, routes data and control to other cells in the array, communicates with the host, and provides data memory.

The features of the Am29325 and the MVISA design system constrained the design. For example, the Am29325's clock rate of 10 MHz determined the timing requirements for the FPC. The Am29325 could have its I/O configured in three different ways, which dictated the structure of the FPC/FPP interface; three 16-bit ports were used for this design. Other factors enabled the design to be less complex, yet maintain high performance.

A preliminary data path design was designed within the constraints. It consisted of two data streams (A and B), three RAMs, three registers to interface with the Am29325 (FF, RR, and SS), and an input port (C-IN). What had not yet been determined was the routing and control requirements. This was done using the simulator.

4.3. FPC/FPP Simulator

A simulator was designed in order to develop the cell design and to prove the design's effectiveness over a broad range of systolic algorithms. It allows verification of algorithms and architectures simultaneously. The utility of various architectural features can be determined, and on the basis of these results, changes can be made to either the algorithm or the architecture.

Developing systolic algorithms for simulation requires several steps. The procedure used here is to first develop a sequential algorithm to generate correct results for comparison with the systolic algorithm. Then, an algorithm is devised for a single cell. Its operation is then verified, assuming for the moment that the inter-cell timing is correct. Finally, the data and address stream delays are verified for the parallel version. The process for developing algorithms was described in section three, where programming systolic arrays was broken up into several modules. The simulation of the array executing an algorithm requires the verification of many different events each cycle. This is especially difficult in the initial stages of the design when both the algorithm and the architecture need to be verified. Memory addresses and data must be specified. The exact timing and data for the host/FPC interface must also be specified. The FPC microprogram must be given in its machine language.

Given the preliminary architecture and a simulator, the data routing and control functions for the cell can be determined using the simulator. These details are omitted here.

Once the register-transfer level design is complete, the design must be translated into a logic design. The translation is straightforward because the architecture has been specified in the 'C' programming language, and the events which occur during each cycle clearly specify the logic and the timing required. After translating the architecture into logic, the timing of the register transfers must be verified. Verification is facilitated by the work done in the previous design stage. The files used to set up the address, data, and control streams can be used unmodified by the MVISA logic simulator. The file which contains the program can also be used unmodified. Cycle by cycle status from the register-transfer level simulations can be used to verify results obtained in the more detailed logic simulations. Other factors simplify the process as well: the design of only one cell must be considered, the register-transfer level design is fixed, and the algorithms have already been verified. The final FPC design is described in the next section.

4.4. Conclusion

In the course of designing two generations of programmable systolic arrays, the importance of good design tools has been evident. Although chip-level design tools were available, it was found that tools to simulate the design at the array level were also needed. To that end, a simulator that could be used in conjunction with design tools was designed in order to develop and verify systolic array algorithms and architectures. After developing an initial design based on a targeted set of algorithms and the constraints imposed by the hardware, the simulator was used to determine an optimized design for the cell's control logic and data routing.

A major problem in designing systolic arrays is the difficulty of verifying the many data transfers that occur during each cycle. To better handle this complexity the programming was divided into modules such as data streams, the math chip program, etc. Also, the data in the simulations are presented in a fashion which promotes easier visualization of the array's operation. Future research directions include verification tools and visualization tools for design and simulation

5. THE FLOATING-POINT PROCESSOR CONTROLLER CHIP: FPC

5.1. Introduction

This section describes in more detail the 32-bit floating-point systolic array cell described in the previous section. The cell consists of a controller chip, the Floating-point Processor Controller (FPC), and a math chip, the the Advanced Micro Devices Am29325 32-bit floating-point processor (FPP) [16]. Following the description of the architecture, a matrix multiplication example is used to demonstrate the cell's unique features.

5.2. FPC Architecture

As shown in Figure 4, with three registers (F, R, S) and two feedback paths, the Am29325 provides in two steps the multiply/accumulate operation -- a key capability needed in many signal processing algorithms. The programmable FPC controls the Am29325, routes data to and from the Am29325, routes data (A and B), addresses (ADDR A and B) and control to other cells in the array, interfaces with the host, and provides local memory for a cell. The FPC was designed for a chip with an 8.4 x 8.4 mm die, 178 pins, and uses a 2-micron NMOS technology [14]. The chip design has passed all logical and physical design checks, but has not been fabricated. The FPC/FPP cell has all the features outlined in [8].

Bit-parallel multiply-accumulate

The Am29325 provides 32-bit floating-point arithmetic at 10 MFLOPS. Using a single-chip FPP reduces the complexity of the cell, and programming is simplified because the Am29325 has no pipelining.

On-chip data RAM

The FPC has two 64 by 32-bit static RAMS for data (RAM A and RAM C). Each RAM may be used once each cycle. As described in section 4, RAM A and RAM C can be swapped. For addresses, each RAM has its own programmable counter. The address control instruction, described later, is used to manipulate the counters.

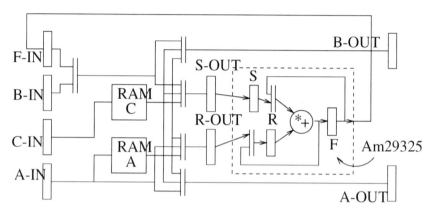

Figure 4. Block Diagram of the FPC/FPP cell.

High-bandwidth inter-cell communication

For data, there are two data streams (A and B) and an extra input port (C-IN). For control, there are three control streams (SA, SB, SC), and a pin for host control (SH). The C-IN data port can be configured as a global port that is loaded by the host and controlled in a daisy chain fashion. Alternatively, C-IN can be used as a third local input port that is loaded by an adjacent cell. This latter arrangement allows the cells to be configured into a hexagonal array.

High-bandwidth intra-cell communication

The FPC has a variety of routing options. Three 32-bit 2-1 multiplexers route data from the four input ports and four 32-bit, 4-1 multiplexers route data to the four output ports. The two RAMs may receive data from ports A-IN and C-IN. There are paths for the A and B streams and for Am29325 routing. F-IN can take results from the Am29325 and load them back into the Am29325 via S-OUT, or route them off-cell via B-OUT.

Control streams provide the communication to synchronize systolic operations. Three systolic control streams are provided, one for each data stream and one for C-IN. Systolic streams SA and SB are one-bit streams which can be programmed to have two to five registers. SB is not shown; it is identical to SA. The first bit is loaded every cycle to allow monitoring of incoming control. The bit can be shifted into the 4-bit shift register by the opcode. SC is a control stream with two registers per cell; it may be used to synchronize RAM C read/writes. Systolic port SH provides a direct connection to the host interface from the FPC. Two bits loaded each cycle into SH indicate one of four modes: scan mode for testing and programming the cell, reset mode, and two normal modes which allow the loading of C-IN to be controlled by the host or by another cell.

Microprogrammed Control

With microprogrammed control, the FPC has complete on-chip parallelism; the Am29325, all data registers, and all data routing are completely controlled for all instructions. The on-chip, 64 x 32-bit program RAM and its simple microsequencer eliminate the need to pass instructions, which reduces I/O. The microsequencer consists of a 6-bit program counter, a 6-bit 4-1 multiplexer, and logic for control. The address for a branch comes from the opcode, or one of the address streams.

5.3. Matrix Multiplication

Simulations have been performed to verify the operation of the FPC/FPP cell. Here, matrix multiplication on a three-cell linear array is described. It is a good algorithm to examine because the data flow and arithmetic operations are similar to operations required in other algorithms.

In this example two 3 by 3 matrices, A and B, are multiplied together to form C, another 3 by 3 matrix ($C = B \times A$). The operands and results are assumed to be in a 32-bit floating-point format. A column of A is loaded into each cell's RAM A. B is fed row by row on the A data stream. With this arrangement, the matrix multiplication is done as a series of vector-vector multiplications. As the results of the multiplication are available, they are fed out on the B data stream. While the multiplication of A is occurring, the next matrix will be loaded into RAM C. To clarify the description of the program, all loops will be unraveled and all don't care operations are not mentioned.

The FPP's vector-vector multiplication program is shown below.

Step	FPP operation			Comment
0.		R = A-IN	S = RAM A	$R = b_{i1}, S = a_{1j}$
1.	F = R * S			$F = b_{i1} * a_{1j}$
2.		R = A-IN	S = RAM A	$R = b_{i2}, S = a_{2j}$
3.		R = R * S		$R = b_{i2} * a_{2j}$
4.	F = R + F	R = A-IN	S = RAM A	$F = b_{i1}a_{1j} + b_{i2}a_{2j}$
				$R = b_{i3}, S = a_{3j}$
5.		R = R * S		$R = b_{i3} * a_{3j}$
6.	F = R + F			$F = c_{ij}$

In this FPP program, $c_{ij} = b_{i1}a_{1j} + b_{i2}a_{2j} + b_{i3}a_{3j}$ is executed in seven steps. The vector products can be pipelined so that only six instruction are required by combining steps 0 and 6. In the FPP, register F stores the accumulated result, S holds A's elements from RAM A, and R holds B's elements from A-IN and stores temporary results.

From the FPP vector product program, the data flow requirements can be determined. In the FPP program, it can be seen that every other cycle (0, 2, 4) R and S are loaded. To feed R, an element from the A data stream every other cycle requires a *LOAD A-IN* during steps 1, 3, and 5, and *A-OUT = A-IN* during steps 2, 4, and 6. Thus data elements b1, b2, and b3, are passed in a systolic fashion from cell 1 to cell 2 over a period of six cycles. To feed S, an element from RAM A every other cycle requires a *SHIFT DA*, and *READ RAM A* during steps 2, 4, and 6.

The data flow on the address stream is quite regular. Every other cycle the address stream contents are shifted and RAM A is read. To read RAM, the address must be available one cycle sooner. To do this the address counter A is initialized at the beginning of the program by an *address* instruction to load every cycle. Thus, an address loaded into AD-IN during step 2 will be loaded into address counter A during step 3, and will be used to read RAM A during step 4.

The final data flow requirement is unloading results. The B stream is used to unload results. Its program is not as regular because one result is output every cycle at the end of each vector product. This is done to speed up the program. To do this, the data is sent faster than the delay between the cells. Like the data flows shown earlier, the delay between cells is three. However, the data stays in each cell only two cycles instead of three, so that data received by cell 2 during instruction i will be received two instructions later by cell 3 during its instruction $i-1$, whereas on the A stream (where data stays in the cell for three cycles) data received by cell 2 during instruction i will be received three cycles later by cell 3 during its instruction i. In the program shown below the step numbers correspond with previous macros, but step 6 is shown first because it is the first event of the unloading process. Nothing is done during steps 4 and 5.

Step	Action	
6.	Load B-IN	
1.	Load B-IN	B-OUT = B-IN
2.		B-OUT = B-IN
3.		B-OUT = F-IN

The program loads a new matrix into a RAM via the global port during each pass through the program. So, the first pass through the program is used to load the RAM with a matrix. To load data via a global port the cell has to be configured into the global mode. Two instructions are needed at the beginning of the program to set the state of the cells. This enables C-IN, RAM C, and SC to be controlled by the host. With host/PE set to host, C-IN will be loaded every cycle, SC will be controlled by SH0, and RAM C will be controlled by SC-IN and bit 21 of the opcode. The cells are loaded loaded with a column of data from A from the left to the right, one by one by sending a 1 along the SC stream like a token in order to coordinate the global loading. The shifting of SC is controlled by the host; SC shifts whenever SH0 is equal to 0.

The requirements for the initialization of the data stream timing can be determined by backtracking from the FPP data flow requirements; data must be loaded into R-OUT and S-OUT the cycle before R and S are needed. During that cycle, the RAM must be read and data on A-OUT must be ready.

To build a complete program these instruction segments are combined.

5.4. Conclusion

A two-chip systolic array cell with 32-bit floating-point arithmetic for signal processing and matrix computations has been designed. The cell uses the Advanced Micro Devices Am29325 floating-point processor (FPP) and a custom controller chip, the Floating-point Processor Controller (FPC) chip. The Am29325 performs all arithmetic, providing 32-bit floating-point multiplication, addition, and subtraction at 10 MFLOPS. The programmable FPC is a custom designed, application specific integrated circuit. It controls the Am29325, routes data to and from the Am29325, and routes data and control to other cells in the array. It has two data memories, one program memory, two data streams, an extra input port, and three systolic control streams. A microsequencer allows concurrent use of all cell resources. The cell has several unique features which provide high performance: 1) use of control registers to reduce the opcode word length, 2) two interchangeable RAMs to enable data to be loaded and used using the same instructions, 3) an input port which can be used either as a global or local port to allow the cell to be used in different array configurations. In addition, the data and control flow architecture is very flexible.

6. CONCLUSION

This chapter examined an approach to the design and programming of systolic array cells for signal processing. The chapter described a cell design and a programming approach. Then, a design approach and a second cell design was described.

Section two described an inexpensive and flexible programmable systolic array cell for signal processing applications. The cell uses two chips: the 16-bit NCR45CM16 CMOS Multiplier/Accumulator (MAC) for arithmetic, and the Systolic Array Controller (SAC), for routing data and controlling the MAC. The SAC has a 64 by 18 bit static RAM which is used twice each cycle, once to read a control word and once to read or write a data word. The SAC has two 16-bit data streams and one 6-bit address stream. A 16-bit bidirectional port routes data between the 72-pin SAC and the 24-pin MAC. All major cell resources can operate concurrently.

The goal was to design an inexpensive and flexible cell. Thus, the cell uses a minimum of hardware -- five data registers, one RAM, one address stream -- in order to minimize the cost of the cell, but the cell uses a variety of techniques in order to achieve a flexible design. The techniques used included limiting routing options, fixing in hardware certain functions, using many resources twice each cycle, using the RAM for programs and data, and using the address stream to carry control and address information.

Section three described a framework for developing systolic array programs. The framework was developed using the SAC/MAC cell. A step-by-step approach for programming the SAC was presented. A key element was the development of a library of macros which implement commonly used program segments. A desired systolic algorithm can be coded by combining several of these macros. In most cases, the programs can be written with little new microcode. With the macro-based approach, the software development steps are programming the MAC, loading RAM, synchronizing cells, delaying data, unloading results, combining the macros into a program, and pipelining a program.

The fourth section examined how to use a simulator to interactively design systolic arrays. From the algorithms targeted for implementation a set of hardware constraints and requirements can be derived, which allows a preliminary architecture to be defined. Hardware resources such as math chips, controllers, and memory are then selected. These hardware resources impose constraints on the design. Then, the desired architecture is implemented using the selected hardware. By using a simulator, detailed results on the performance of the cell are obtained which help the designer modify the array and the architecture to achieve the desired performance. After a series of iterations the systolic array design is complete, and translation into hardware progresses in a straightforward manner.

Section five presented the FPC/FPP cell, a second systolic array cell design, which uses floating-point math chip and has a more flexible controller chip than the SAC/MAC cell. To achieve a flexible and efficient design the architecture includes on-chip RAM, wide communication paths, bit-parallel multiply/accumulate, parallel control, small number of chips, and microprogrammed control.

Future work in this area could involve using the simulator to examine other architectural options, and the tools could be modified to include additional features to aid the designer.

REFERENCES

[1] H. T. Kung and C. E. Leiserson, "Systolic Arrays (for VLSI)," *Sparse Matrix Proceedings, 1978*, Academic Press, Orlando, 256-282, 1979; also in "Algorithms for VLSI Processor Arrays," in *Introduction to VLSI Systems*, C. A. Mead and L. A. Conway, eds., Addison-Wesley, Reading MA, pp. 271-292, 1980.

[2] H. T. Kung, "Why Systolic Architectures?," *Computer*, pp. 37-46, Jan. 1982.

[3] D. E. Foulser and R. Schreiber, "The Saxpy Matrix-1: A General-Purpose Systolic Computer," *Computer*, pp. 35-43, July, 1987.

[4] J. V. McCanny and J. G. McWhirter, "Some Systolic Array Developments in the United Kingdom," *Computer*, pp. 35-43, July, 1987.

[5] M. Annaratone et al., "Architecture of Warp," *Proceedings COMPCON*, pp. 264-267, Spring, 1987.

[6] M. Annaratone, E. Arnould, T. Gross, H. T. Kung, M. Lam, O. Menzilcioglu, and J. A. Webb, "The Warp Computer: Architecture, Implementation and Performance," *IEEE Transactions on Computers*, C-36, Vol. 12, pp. 1523-1538, December 1987.

[7] P. G. Ducksbury, *Parallel Array Processing*, Ellis Horwood Limited, Chichester, West Sussex, England, 1986.

[8] A. L. Fisher, H. T. Kung, L. M. Monier, and Y. Dohi, "Design of the PSC; A Programmable Systolic Chip," *Proc. of the 3rd Caltech Conf. of VLSI*, pp. 287-302, March, 1983.

[9] G. Nudd, "Concurrent Systems for Image Analysis," *VLSI for Pattern Recognition and Image Processing*, Ed. K. Fu., pp. 107-132, 1984.

[10] P. J. Kuekes and M. S. Schlansker, "A One-third Gigaflop Systolic Linear Algebra Processor," *SPIE Vol. 495 Real Time Signal Processing VII*, pp. 137-139, 1984.

[11] A. L. Fisher, *Implementation Issues for Algorithmic VLSI Processor Arrays*, Ph.D. thesis, Carnegie-Mellon University, Dept. of Computer Science, Oct., 1984.

[12] F. H. Hsu, H. T. Kung, T. Nishizawa, and A. Sussman, "LINC: The Link and Interconnection Chip," CMU-CS-84-159, Carnegie Mellon University, 1984.

[13] NCR Corporation, "NCR45CM16 CMOS 16 X 16 bit Single Port Multiplier/Accumulator," product brochure, 1984.

[14] W. H. Elder, P. P. Zenewitz, and R. R. Alvavordiaz, "An Interactive System for VLSI Chip Physical Design," *IBM Journal of Research and Development*, pp. 524-536, Sept., 1984.

[15] E. J. McCluskey, "Design for Testability," Chapter 2 of *Fault Tolerant Computing; Theory and Techniques, Vol. 1*, Ed. D. K. Pradhan, Prentice-Hall, Englewood Cliffs, 1986.

[16] R. Perlman, "A New Approach to Floating-Point Digital Signal Processing," *VLSI Signal Processing*, P. R. Cappello et al., editors, pp. 400-410, 1984.

[17] B. Friedlander, "Block Processing on a Programmable Systolic Array," *Proceedings COMPCON*, pp. 184-187, Spring, 1987.

[18] H. T. Kung and M. S. Lam, "Wafer-Scale Integration and Two-Level Pipelined Implementation of Systolic Arrays," *Journal of Parallel and Distributed Computing 1*, pp. 32-63, 1984.

[19] L. Snyder, "Programming Environments for Systolic Arrays," *Proceedings SPIE 614, Highly Parallel Signal Processing Architectures*, K. Bromley, editor, pp. 134-143, 1986.

[20] J. A. Webb and T. Kanade, "Vision on a Systolic Array Machine," from *Evaluation of Multicomputers for Image Processing*, Eds. Uhr, Preston, et al. Academic Press, 1986.

[21] B. Bruegge et al., "Programming Warp," *Proceedings COMPCON*, pp. 268-271, Spring, 1987.

[22] H. T. Kung, "Systolic Algorithms for the CMU Warp Processor," *Proc of the 7th Intl. Conf. on Pattern Recognition*, pp. 570-577, July, 1984.

[23] M. Annaratone et al., "Applications and Algorithm Partitioning on Warp," *Proceedings COMPCON*, pp. 272-275, Spring, 1987.

[24] L. Melkemi and M. Tchuente, "Complexity of Matrix Product on a Class of Orthogonally Connected Systolic Arrays," *IEEE Trans. on Computers*, Vol. C-36, No. 5, pp. 615-619, May, 1987.

[25] H. V. Jagadish, S. K. Rao, and T. Kailath, "Array Architectures for Iterative Algorithms," *Proceedings of the IEEE*, Vol. 75, No. 9, pp. 1304-1321, Sept., 1987.

[26] R. Smith, M. Dillon, and G. Sobelman, "Design and Programming of a Flexible, Cost-Effective Systolic Array Cell for Digital Signal Processing," *IEEE Transactions on Acoustics, Speech and Signal Processing*, Vol. 38, No. 7, pp. 1198-1210, July 1990.

[27] J. M. Speiser, "Linear Algebra Algorithms for Matrix-Based Signal Processing," *Proceedings SPIE 614, Highly Parallel Signal Processing Architectures*, K. Bromley, editor, pp. 2-12, 1986.

[28] G.-J. Li and B. W. Wah, "The Design of Optimal Systolic Arrays," *IEEE Transactions on Computers*, Vol. C-34, No. 1, pp. 66-77, 1985.

[29] D. I. Moldovan and J. A. B. Fortés, "Partitioning and Mapping Algorithms into Fixed Size Systolic Arrays," *IEEE Transactions on Computers*, Vol. C-35, No. 1, pp. 1-12, Jan. 1986.

[30] R. Smith and G. Sobelman, "Simulation-Based Design of Programmable Systolic Arrays," *Computer-Aided Design*, June 1991.

7

Analog VLSI Signal Processors:

Design and Test Methodologies

Mani Soma
Design, Test & Reliability Laboratory
Department of Electrical Engineering, FT-10
University of Washington
Seattle, Washington 98195
e-mail: soma@ee.washington.edu

Abstract

Analog signal processors have received increasing attention in the past few years thanks to the advances in fabrication technologies permitting better control of device characteristics and to the explosive growth in neural networks. While digital processors still dominate design applications, new problems in neural signal processing, pattern recognition, and image processing lend themselves quite nicely to analog implementations, and quite a few case studies, especially those in neural networks, have demonstrated the superiority of analog signal processing circuits and systems.

This chapter focuses on the design and test methodologies for analog VLSI signal processors. Recent advances in circuit design techniques and new circuit configurations will be covered. The impact of better fabrication processes, especially for MOS devices, is a remarkable improvement in both circuit performance and complexity. The complexity issues surface more clearly in the verification of processor functions and performance. Research results from a new discipline, design-for-test (DFT) techniques for analog VLSI systems, will be presented together with examples illustrating their applications in testing analog signal processing subsystems. These DFT techniques are incorporated into a comprehensive design methodology that translates a specific algorithm to circuits suitable for IC realization.

I. INTRODUCTION

Signal processing, especially since the late 1970's, has almost been synonymous with *digital* signal processing (DSP) due to the advantages of digital designs: low-noise, very high accuracy, high speed, advanced process technologies especially for Complementary Metal-Oxide Semiconductor (CMOS) VLSI systems, high-yield manufacturing, etc. Analog designs, meanwhile, evolve at a much slower pace and derive some fundamental benefits from the digital advances. The improvement in process technologies spills over to the analog domain to help the designs of high-precision low-noise circuits using small-geometry CMOS devices and, more recently, BiCMOS (bipolar-CMOS) devices. Circuit density also rises at a relatively fast rate and analog circuits approach LSI subsystems.

The two factors that finally bring analog designs to the leading edge are neural computing and, ironically enough, very high-speed digital computing. Neural networks, as reviewed below with respect to analog signal processors, have been implemented using "imprecise" analog circuits and, as claimed by several authors, achieve much better performance than their digital counterparts. Very high-speed digital systems, in which clock frequencies exceed 100 MHz and rise times are of the order of a few nanoseconds, resurrect the classical problems of noise and waveform degradations, and digital designers suddenly find themselves go back to basics to deal with these analog predicaments. These two factors will be reviewed in this chapter to set the stage for the discussion of analog design techniques and of the requirements necessary to make analog systems truly VLSI. These requirements encompass advances in process technologies, design tools, and test tools, which contribute towards a comprehensive design methodology.

II. BACKGROUND

1. Analog techniques in neural computing

The history of neural computing and recent developments is well reviewed by Hecht-Nielsen [1], and this section focuses only on analog neural processors. Mead [2] in his lucid text on analog VLSI argues convincingly that analog design techniques are better suited to realize many functions performed by the neural systems. Using subthreshold MOS circuits as the building blocks (to be explained in more details below), several systems reproducing sensory functions (optical retina, cochlea) were designed and shown to perform satisfactorily. Numerous other works in this area follow [3-7] and while neural computing systems are still predominantly digital [1], analog circuit techniques for this specific application have gained a firm foothold with well defined methodologies [6] and will remain a fundamental component of the overall analog VLSI signal processing systems. The major impact of neural computing on analog designs is undeniable: interest in analog designs surges and system implementers re-discover the wealth of functions realizable without resorting to digital techniques, which sometimes may be more costly due to the large word size and other complexities. The combination of analog functions with the existing digital system design methodologies produces numerous hybrid signal processors, which have become the norm in neural computing.

2. High-frequency digital systems

Submicron CMOS process technologies have resulted in very high speed digital systems with rise or fall times of the order of a few nanoseconds. Gallium arsenide (GaAs) or bipolar digital subsystems achieve even faster clock speeds and transition times. The digital signals in these systems are, interestingly enough, no longer "square." A 500 MHz clock, on an oscilloscope display, in fact resembles a distorted sine wave. Frequency effects usually occurring in analog and microwave circuits (e.g. skin effects, crosstalk, substrate noise mechanisms, etc) become intrinsic in digital system designs, and classic analog strategies dealing with these difficulties are being re-discovered. Another consequence of advanced process technologies and smaller device geometries is the reduction in the power supply level from 5V to 3.3V to guarantee non-breakdown and / or low-power operations. This reduction adversely affects the noise margin of digital circuits and in fact makes these circuits as much susceptible to noise as their analog cousins. The only difference between the two circuit types seems to be that digital signals are sampled and used only at the high and low voltages while analog signals use all the values. Since "high" and "low" voltages are increasingly closer to each other due to shrinking power supplies, noise contamination and computation accuracy will be the major predicaments for the digital designers in the near future unless analog techniques are sufficiently improved. It can be argued thus that at high speed and low voltages, which is the trend in digital VLSI designs, *all* signal processors are analog signal processors.

III. ANALOG CIRCUITS AND SUBSYSTEMS FOR SIGNAL PROCESSING

The wealth of analog circuits and subsystems is unparallel and has been described in details in numerous textbooks [8-10] and collections of research papers [11]. The discipline is very mature despite the usual misconception that analog design is an art rather than a science. Any summary of analog subsystems for signal processing is necessarily incomplete but we will attempt to highlight the four design techniques more pertinent to the current state-of-the-art in analog VLSI: continuous-time designs, switched-capacitor (SC) designs, switched-current (SI) designs, and subthreshold designs. The first two methodologies are well established and are the cornerstones of current signal processors while the last two methodologies still evolve at a very fast pace with strong potential to make purely analog VLSI a reality.

1. Continuous-time circuits

The traditional approach to circuit design produces two large classes of continuous-time circuits: passive analog and active analog circuits. The theory of passive circuits is fully developed and culminates in well defined methodologies for network synthesis, automatic filter synthesis, and control system designs. However, since passive circuits do not play a significant role in large-scale signal processors, we will concentrate on active analog circuits. The classic text on analog

IC analysis and design by Gray and Meyer [8] contains a lucid description of the major bipolar circuits and subsystems. MOS circuits, relatively neglected in this text, receive full treatment in other works [9-11] and serve as the fundamental building blocks for larger subsystems, especially with regard to filter applications. One essential characteristic of continuous-time active circuits deserves to be mentioned: the principle of device matching. The current mirror and the differential amplifier in figure 1 illustrate this principle: the matching between M1 and M2 of the current mirror ensures a stable current source and the matching between M3 and M4 of the differential pair ensures the common mode rejection. Almost all successful analog designs exploit device matching and relative scaling (i.e. one device larger than the matched device by a designed scale factor). Since exact match is experimentally impossible, any mismatch creates signal offset, which is a major error source in active circuits. Other error sources include device parasitics (e.g. diode capacitances at the source or drain junctions of MOS devices), leakage currents, distortion in a linear design, etc. In short, these are the classic errors that plague analog circuits and make them unattractive for system applications.

The climax of continuous-time analog circuits is the ubiquitous operational amplifier (opamp for short), one of whose MOS implementations is depicted in figure 2. The circuit has a large input impedance, a reasonably large voltage gain, and a relatively low output impedance. These characteristics, common to most opamps in either bipolar or MOS technologies, make the opamp the fundamental building block in subsystem designs: general-purpose amplifiers, multipliers, filters, comparators, digital-to-analog converters (DAC), analog-to-digital converters (ADC), oscillators, phase-locked loops, etc. Continuous-time techniques, while producing the popular opamps, fail to cover the complete design of these subsystems due to the requirement of passive components (resistors and capacitors) to realize the designs. These passive components either take up massive space on the IC or are provided off-chip, thus increasing manufacturing cost and packaging complexity. Since their values are also quite difficult to control precisely, they are useful mostly for small-scale signal filter designs where accuracy, either in voltage level or in timing, is not crucial.

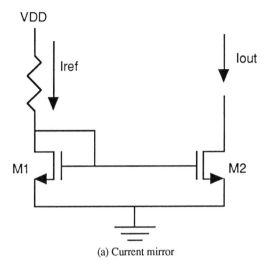

(a) Current mirror

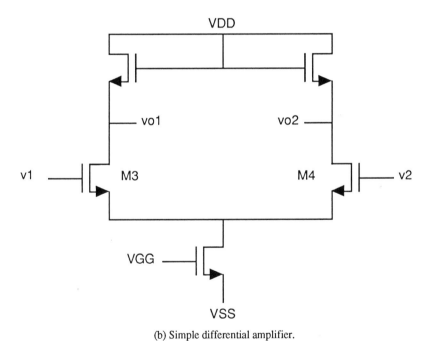

(b) Simple differential amplifier.

Figure 1. (a) Current mirror, (b) Simple differential amplifier.

190

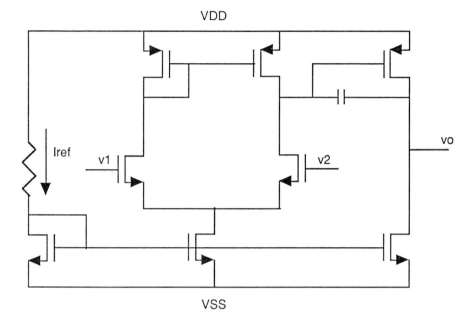

Figure 2. Simple CMOS operational amplifier.

2. Switched-capacitor (SC) circuits

SC circuits ameliorate one aspect of classical continuous-time circuits: precise control of component values for capacitors and "resistors." Resistors actually do not physically exist but are implemented by combinations of capacitors, MOS switches, and control signals (figure 3). This implementation is not new [12-13] but it was relatively dormant until 1970's. The technique received a forward thrust in the 1970's from the advances in MOS technologies and has turned out to be quite suitable for integration with digital systems since it relies on sampled data methodologies to process analog signals. The replacement of resistors by capacitors, as illustrated in figure 3, is valid only during specific clock phases and is not exact in most cases. However, SC circuits are sufficiently powerful to handle most signal processing tasks and are currently the most dominant proven methodology in analog designs. The precise control of the capacitors is accomplished by laying out each capacitor as a multiple of a unit capacitor, thus the ratio of capacitances becomes the ratio of the number of units, which suffers only very small errors (0.1% or less) in manufacturing. The precise control of resistors is accomplished via this technique and the accurate setting of timing pulse width, a fact which has already been accomplished in digital systems.

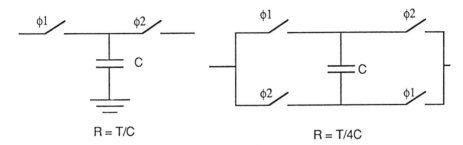

Figure 3. Examples of resistor realizations in SC circuits.

SC circuits also benefit in terms of layout size with the absence of resistors. The clock phases merge smoothly with digital clocks, and thus analog sampled data processing integrates seamlessly with digital signal processing. SC methodologies are quite mature [9-11] and have been used to design subsystems covering the entire analog domain: filters (its most important application to date), comparators, nonlinear circuits, oscillators, modulators, phase-shifters, phase-locked loops, DAC, ADC, and programmable SC circuits. The major drawbacks stem from the analog aspects of the design: device parasitics, noise, nonlinearity, significant layout size of capacitors. Frequency performance is sometimes limited by the fundamental assumption that all capacitive charging and discharging must reach the steady state as soon as possible during each clock pulse. Given specific capacitor values dictated by the design, e.g. by a filter, the clock frequency has to be adjusted to satisfy this assumption and usually is in the range of 100 KHz, thus is not as fast as desired in many processing functions.

3. Switched-current (SI) circuits

SI circuits take another step toward the elimination of passive devices in circuit design by getting rid of the capacitors used in SC circuits. While sampled data methodologies are still the fundamental tenet of SI signal processing, the signal is now measured in terms of currents instead of voltages, and this simple change of signal variable turns out to be quite significant since it permits complete analog circuit designs using only active devices. From another perspective, the charge equation:

$$Q = C\,V = I\,t \qquad (1)$$

can be manipulated in two different ways. SC circuits choose to deal with signal as voltages, thus must provide precise capacitors to manipulate these signals. SI circuits make the second choice by dealing with signals as currents, thus have to provide precise timing controls to manipulate these signals. Since precise timing is easy to accomplish using only active devices as shown by digital designs, SI techniques eliminate passive devices and potentially bring analog technologies to true VLSI level.

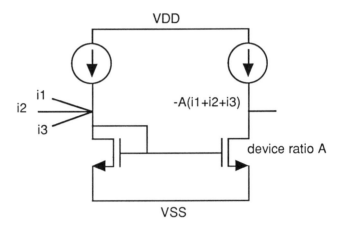

(a) Summation and scaling [14].

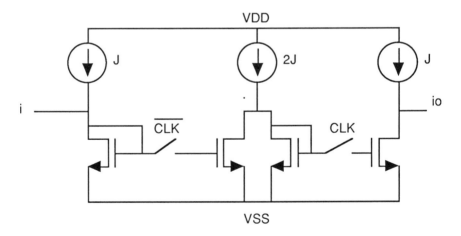

(b) Delay circuit [14].

Figure 4. Switched-current circuits.

The basic operations needed for signal processing are scaling, summation, inversion, and delay. Using the simple current mirror as the basic block to illustrate SI designs, circuits accomplishing these tasks [14] are shown in figure 4. Scaling is built in by ratios of device sizes (figure 4a), summation is performed by the current mirror (figure 4a), delay is easily incorporated by cascading current mirrors with clock-controlled switches (figure 4b), inversion is designed using differential current mirrors. A design of a differential integrator is shown in figure 4c [15] and illustrates SI techniques at subsystem level. It should be noted that these examples built on current mirrors are chosen only to describe the concepts of SI designs. Better circuits, e.g. using cascode transistors, are required in realistic designs to meet performance specifications but their principle of operations is identical to these simple circuits.

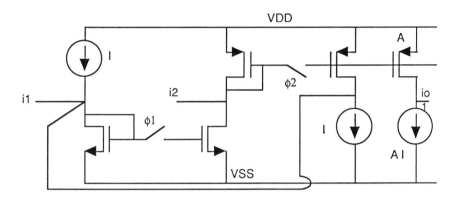

(c) Differential integrator [15].

Figure 4. Switched-current circuits (continued).

SI signal processing systems can be built based on these basic modules for the following reasons [14]:

1. Device matching is required only within a module, and each module is self-contained.

2. Only signal currents are transmitted between modules.

3. Interconnection between modules results in little loss of accuracy due to impedance mismatch between the driving module (high output impedance) and the input stage of the next module (low input impedance).

SI circuits, while representing a significant step forward in analog VLSI signal processing, suffer from device mismatch effects and more importantly, current feedthrough due to the switching clocks (figure 4b). Fully differential designs and techniques derived in SC circuits can be used to minimize feedthrough [16], common mode signals, and power supply fluctuations. Error cancellation in SI designs is a subject of intensive research at this time.

While SI design methodologies show strong promises, the discipline is still young and growing. SI filters [15] and converters [17-18] have been designed and tested and their performance is comparable to or exceeds that of SC modules. No system application has been attempted however, but the principle underlying SI designs does not seem to contain any intrinsic impediment and should be easily extensible to VLSI systems.

4. Subthreshold analog circuits

Subthreshold design has almost always been an anathema to designers due to the inherent problems with noise. Standard design techniques are already plagued by noise and inaccuracies, and subthreshold operation seems to worsen the situation. However, Mead [2] shows that subthreshold MOS circuits do perform valuable functions and several major analog VLSI signal processors [3-7] in the area of neural networks have been designed and tested based on this design discipline. The circuit topologies are not new: they are identical to normal MOS circuits in most cases. The critical difference is in the circuit operation: instead of devices operating in linear or saturation regions, they are actually in subthreshold regions where the drain current is an exponential function of gate and drain voltages. The exponential characteristic, similar to the bipolar devices, guarantees high transconductance and also makes it possible to design subthreshold circuits by mimicking bipolar circuits. Large power losses occurring in bipolar circuits are avoided since subthreshold currents are in the nano-ampere range. The design of the basic modules is described in details in [2] and a library of circuits for neural applications is discussed in [6]. Within the context of neural networks, the issue of noise is resolved based on the intrinsic fault tolerant properties of neural circuits with a very large number of inputs, self-adjusting properties to change weights to reduce errors, and learning properties. In a sense, these systems adapt around their own errors, thus even though each small module might be contaminated with noise, the overall system still performs satisfactorily. This interesting concept of building a "perfect" system out of "imperfect" components is not new in electronic designs (the classic differential pair is an example) but is manifested much more prominently in neural and other biological computing systems. With regard to subthreshold designs as well as other design techniques mentioned above, this concept has to be exploited to the fullest if analog VLSI signal processors are to compete successfully with the existing digital systems.

IV. ANALOG VLSI SIGNAL PROCESSORS: CURRENT STATE AND DESIGN REQUIREMENTS

The description of the four major analog design disciplines in the above section reveals that the two techniques with strong potential for VLSI are SI designs and subthreshold designs. Analog VLSI has already been demonstrated using subthreshold methods but these methods have not been extended to the usual signal processors. SI methods are still currently at the subsystem level even though some system applications should be forthcoming. The breakthrough from LSI to VLSI, just like in the digital case in the late 1970's, requires not only new circuit techniques but also computer-aided design (CAD) tools. Without these tools, system design, especially in analog, is an impossibility. We have covered circuit techniques in reasonable details in this chapter, and will now present a discussion on the computer-aided design tools necessary to make analog VLSI practical and accessible.

1. Design and synthesis

Existing tools for analog designs focus on circuits and subsystems. Within the framework of application-specific ICs (ASICs), most foundry vendors and system houses offer analog cells or gate arrays integrable with digital cells on the same IC. The gate arrays contain discrete devices and isolated circuits (current mirrors, opamps, etc) to be connected by users to perform desired functions. Standard analog cells emphasize opamp-based modules (e.g. simple amplifiers, differentiators, etc) with predetermined performance. These design techniques force the designer to comply with existing circuits since modification is either difficult or impossible. The underlying principle in these gate arrays and cell libraries is that analog circuits are used only as a small part of a hybrid IC: input / output functions are analog but signal processing functions are digital. Since most hybrid signal processors at this time fit quite well into this architecture, the standard cell and gate array approaches are sufficient to meet the design requirements.

Analog VLSI, on the other hand, will require tools much more sophisticated than just a cell library. The drive to develop new tool sets for analog design currently takes place in research and academic laboratories since there is no truly analog VLSI products to justify large industrial investments. With respect to analog synthesis, the major effort is the OASYS project at Carnegie Mellon University [19]. This project provides a hierarchical structured framework for analog circuit synthesis with two important features: it decomposes the design task into a sequence of smaller tasks with uniform structure, and it simplifies the reuse of design knowledge. Mechanisms are described to select from among alternate design styles, and to translate performance specifications from one level in the hierarchy to the next lower, more concrete level. OASYS has been used to synthesize successfully subcircuits such as filters and is being extended to cover other circuit classes.

A prototype compiler, marketed by Seattle Silicon Corporation (SSC) between 1983 and 1988, automatically generated opamps and filters based on performance specifications and were used quite extensively at the University of Washington, but due to the lack of industry interest, the compiler development effort was discontinued by SSC in 1988. More recent efforts to develop performance-driven layout tools have taken place at various laboratories but these works are still too early to provide a significant impetus to propel analog VLSI development.

2. Simulation

The major obstacle to analog designs is simulation. The classic SPICE and its popular derivatives (PSPICE, HSPICE, etc) are still the mainstay of analog simulation. PSPICE, HSPICE, and related programs now offer model development tools to help users incorporate their own models, schematic entry, and graphic post-processors to make simulation more convenient and powerful. Some programs (e.g. PSPICE) also provide simulation interfaces with digital logic circuits so mixed-mode simulation is, to some degree, possible. Saber, a product marketed by Analogy, claims to solve the mixed-mode simulation problem and more recently, Analog Artist, marketed by Cadence, also makes the same claims even though both programs have not been exercised extensively to justify the claims. Mixed-mode hierarchical simulation without cumbersome translation (from analog to digital or vice versa) is still a subject of active research and development. MISIM [20], a research simulator based on waveform relaxation algorithms, seems promising but has not provided any concrete evidence of success at the time of this writing.

It should be mentioned that SPICE's major anathema, non-convergence, still haunts some new improved simulators. In most instances, users are to blame since the customized models fail to conserve charges or contain discontinuities, but there are instances where the algorithms are unable to handle the tasks. To address these problems, some efforts have been made to use expert systems to assist simulation but results so far are too scanty or too specific to an application to judge. The issue of long simulation time is also an impediment to further progress. Currently, analog circuits with only thousands of devices may take days to simulate, thus a true analog VLSI circuit with up to millions of devices would not be amenable to any simulation at all.

For analog VLSI to become a reality, simulators must exist to facilitate fast and accurate simulation at all levels: circuit, subsystem, and behavior. While there are current research projects addressing these issues, these efforts seem quite inadequate compared to the size of the problem, and it remains to be seen if the next few years will witness a breakthrough in analog simulation.

3. Test and design-for-test (DFT)

No technologies ever mature to the VLSI level without tools for testability analysis, test generation, and performance verification after manufacturing. This section will review the state-of-the-art in analog and mixed signal test research, and discuss a methodology applicable to continuous-time and SC techniques [21] to set the stage for further research in analog testing.

The design of analog ICs has always been a challenge due to the sensitivity of circuit parameters with respect to component variations and process technologies. To ensure the testability of a design is an even more formidable task since testability is not well defined within the context of analog ICs. While the number of I/O of an analog IC is reasonably small compared to that of a digital VLSI circuit, the complexity due to continuous signal values in the time domain, and the inherent interaction between various circuit parameters impede efficient functional verification and diagnosis. Analog testing frequently relies on acceptable operational margins for critical parameters, and analog diagnosis is still more of theoretical interest rather than experimental applications.

In view of this complexity, the general problem of DFT for analog circuits is almost certainly intractable. However, for specific classes of circuits such as those used in signal processing, numerous attempts [31-33] have been made to improve some measure of testability. Testability is defined as controllability and observability of significant waveforms within the circuit structure. Within the context of filter circuits as an example, the significant waveforms are the input / output (I/O) signals of every stage in the filter, and a DFT methodology should permit full control and observation of these I/O signals from the input to the first stage and the output of the last stage of the filter. We assume of course that the first-stage input is controllable and the last-stage output is observable.

It is clear that the types of failures to be studied have a strong impact on the DFT technique, and thus the fault model will be discussed after previous works on DFT techniques applicable to these faults are reviewed. The fundamental theory underlying the proposed methodology is presented followed by an algorithmic description and the results of several case studies. These results are considered from the perspectives of the overhead incurred and the extension of this technique to other analog structures.

3.1 Fault models.

Numerous works [22-26] have been devoted to testing for catastrophic failures in analog circuits, i.e. failures that manifest themselves in a totally malfunctioned circuit. These failures, frequently called catastrophic failures, are easy to detect but difficult to locate and correct. Other faults result in performance outside specifications, and the problem of these out-of-specification faults is inherently more difficult since the circuit still works but is unacceptable, e.g. a low-pass filter with wider bandwidth than as designed. Various faults within the circuit could give rise to the observed effects at the output (or, to use the prevalent terminology in digital testing, these faults are equivalent), thus fault isolation is an important aspect of testing for these non-catastrophic faults.

The detection of out-of-specification faults or catastrophic faults is fairly easy to accomplish based on a simple comparison between simulated outputs and measured data. Further diagnosis immediately poses several difficult questions: what test waveforms should be applied? do they need to be applied in some sequence? how can the measured data be interpreted to isolate the faults? The first two questions involve test generation techniques, and the third involves fault identification for a given circuit. In the methodology to be discussed below, the faults to be studied are also assumed to be limited to the passive components synthesizing the input and feedback impedances of each filter stage, i.e. the operational amplifiers are fault-free. This assumption may be justified by two reasons:

1. Filters designed on printed-circuit boards employ packaged operational amplifiers, which have been tested and are functional.

2. Filters designed on integrated circuits frequently use operational amplifier cells either generated by a silicon compiler or from a macrocell library. In either case, the cell design has been proven correct. Random faults due to the fabrication process are eliminated by the standard wafer qualification tests, and are not germane here.

Fault modelling of operational amplifiers is a topic in itself, and some of the works cited in the next section may be consulted for further discussion.

Most previous works in the area of analog fault modelling and diagnosis focus on the theoretical aspects of the problem. Rapisarda and DeCarlo [22] uses a tableau approach in conjunction with a Component Connection Model to generate a set of fault diagnosis equations, Lin and Elcherif [23] studies the analog fault dictionary approach. The more theoretical works include Hakimi's extension of the t -fault diagnosable concepts to analog systems [24], Togawa's study of linear algorithms for branch fault diagnosis [25], Huang and Liu's method for diagnosis with tolerance [26], Visvanathan and Sangiovanni-Vincentelli's theory of diagnosability [27-28], Trick and Li's analysis of sensitivity as a diagnostic tool [29], and Salama's decomposition algorithm for fault isolation [30]. These theoretical works rely extensively on the characteristic matrix of the circuit under test to define conditions for testability and diagnosability, and are not directly applicable to analog VLSI systems. Analog test generation and design-for-test techniques have also been studied [31-32]. More recently, several new works were presented at the International Test Conference (1988), which include a testability evaluation tool [33], a test generation algorithm for DC testing [34], and a DFT scheme based on signal multiplexing [35]. These recent works approach the problem of analog testing from a more experimental perspective and discuss some fundamental principles regarding controllability and observability of internal signals. While the traditional techniques based on matrix solution and sensitivity analysis are still prevalent, the multiplexing technique [35] approaches a real DFT methodology where tradeoffs seem reasonable.

3.2 Fundamental theory.

The success of scan techniques in digital testing relies heavily on the register structure in sequential logic designs. A straightforward application of these techniques to analog circuits requires similarity in structures, and one circuit class exhibiting this configuration is a multi-stage active analog filter. The most significant functional difference between the analog filter stage and the register flipflop is that each stage does not store analog information in the same way that a flipflop stores digital information. Instead, the signal is modified by the stage gain and bandwidth before passing on to the next stage. Waveform scanning to control and observe signals within the filter is possible only if somehow the bandwidth limitation can be relaxed to accommodate the test signals for individual stages. The word "scan" is used here only in the figurative sense to demonstrate the philosophical similarity with the digital scan path technique, since literally the signals are continuously transmitted in analog circuits and there is no need for any clock to enable and disable this transmission. Waveform scanning to improve the controllability and observability of signals can be accomplished if the bandwidth of each stage can be dynamically broadened in the test mode. The attendant change in gains is unavoidable but is not a real issue since any gain change, as demonstrated below, involves a fixed scale factor and can be taken into account by programming the test equipment to compensate for this factor. Thus the critical problem here is dynamic bandwidth expansion.

This bandwidth expansion to enable the scanning of waveforms in and out of a filter stage under test is performed by reducing the capacitive effects in the impedances of the stages not under test. Given an active analog filter, all the impedances are realizable based on a combination of four fundamental connections of resistors and capacitors: single-resistor branch, single-capacitor branch, series RC branch, and parallel RC branch. We now describe four canonical transformations [21] which in effect, disconnect the capacitors from the circuits using simple MOS switches. These canonical transformations form the basis for all impedance modifications involved in making the filter testable.

1. Single-resistor transformation.

An ideal resistor has an unlimited frequency bandwidth, thus the transformation in this case is the identity transformation, i.e. the resistor does not need to be modified.

2. Single-capacitor transformation.

The single-capacitor branch cannot simply be disconnected by a single MOS switch since such a disconnect could create noise and instability problems, e.g. when the capacitor is the only feedback element in an integrator subcircuit. Thus the transformation involves two MOS switches as shown in figure 5a. The impedance Z_N is approximately the same as the original impedance Z only if R_S is small enough so that the zero created does not affect the frequency response of the stage or of the overall circuit.

3. Series RC transformations.

The capacitive effect can be reduced by two possible transformations (figure 5b): a switch in series with the capacitor to disconnect the branch in the test mode or a switch in parallel with the capacitor to make the branch resistive in the test mode. To avoid significant perturbations of the original pole-zero locations in the series-switch configuration, R_S must be much less than R.

4. Parallel RC transformations.

Only one switch is sufficient to accomplish the transformations. The added switch can be either in series with the capacitor to disconnect it or in parallel with the capacitor to short it out in the test mode. The two transformations (figure 5c) are analogous to that presented above for the series RC branch. To avoid significant perturbations of the original pole-zero locations in the series-switch configuration, R_S must be much less than R .

3.3 DFT procedure.

The design-for-test (DFT) procedure follows directly from the above discussion of the modification techniques to improve the bandwidths of individual filter stages. Given the design of a conventional active filter with N stages, each stage i has bandwidth B_i and gain G_i . Let $B = max \{B_i , i=1..N\}$. The DFT algorithm is:

1. Consider the first stage ($i =1$). Insert MOS switches according to the above transformations to widen the stage bandwidth to at least B . Define the control waveforms necessary for normal mode and test modes.

2. Repeat step 1 for stages $i = 2 ..N$.

3. Overhead reduction of control lines: from the control signals defined in steps 1 and 2, establish sets of common control lines (i.e. lines that always have identical logic values during test and normal modes). Replace each set of common lines with one single line.

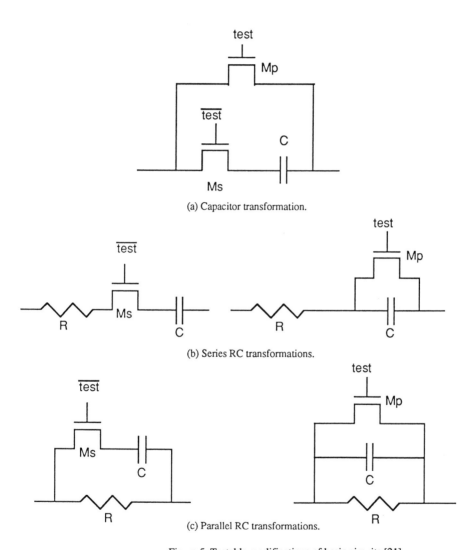

(a) Capacitor transformation.

(b) Series RC transformations.

(c) Parallel RC transformations.

Figure 5. Testable modifications of basic circuits [21].

4. Overhead reduction of MOS sizes: from the overall pole-zero diagram of the modified filter, consider additional movements of the poles / zeroes introduced by the test switches to see if these devices can be made smaller without impeding signal scanning during test mode.

5. Test generation: for each stage under test (SUT), the overall circuit topology and transfer function (from steps 1 and 2) are used to generate the necessary test input waveforms. A circuit simulator, e.g. SPICE, provides the expected outputs (waveforms, gain, phase, etc). Given a filter with N stages, N+2 simulations are required per fault: 1 simulation of the circuit with the fault, N simulations (1 for each SUT) in test mode, and 1 simulation in all-test mode (i.e. all stages are transformed to simple gain stages with all capacitors disconnected by controlling the switches).

6. Fault isolation: the simulated and measured data are interpreted (see case study below) to identify and isolate the fault.

The test application is also straightforward. The circuit is first tested in normal mode. If the circuit fails (either due to a catastrophic fault or out-of-specification fault), the test mode is activated and one stage is tested at a time to isolate the failure. The waveforms to be used are from the test generation step (step 5) above. Fault isolation is then carried out to identify the specific fault and the stage where the fault occurs.

3.4 Case study and results.

The above DFT technique has been applied to numerous case studies [21] including low-pass, band-pass, high-pass, and notch filters. The filter types include Chebyshev, Butterworth, inverse Chebyshev, and elliptic. This section selects one of these circuits and presents an application of the methodology. The results are shown from simulation, even though the circuits themselves have been built and tested. The measured data are within experimental errors of the simulated data, thus only the simulated data are discussed to simplify the presentation. The circuit used for this case study is a combined band-pass and low-pass biquad filter in figure 6 (without the bold-faced MOS devices). The simulated data are obtained from PSPICE using the Motorola MC14573 as the operational amplifier. The filter was also built with this operational amplifier on a circuit board and tested. The filter has 2 zeroes at infinity and 2 poles at $-50 \pm j1000$ radians, corresponding to a pole frequency of 159 Hz.

The modification to enable test modes involves adding 3 MOS switches (figure 6): M_{s1} in series with C_1 (per parallel RC branch transformation), M_{s2} in series and M_{p2} in parallel with C_2

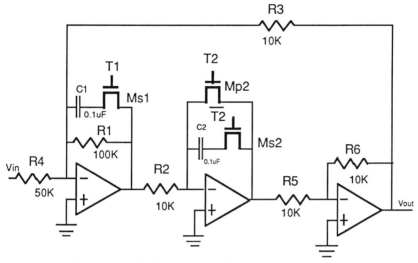

Figure 6. Testable filter case study [21].

(per single-capacitor transformation). The transfer function in normal operation (the 2 series switches are on, the parallel switch is off) is:

$$H_M(s) = \frac{-(R_6/R_2R_4R_5)\,(s^2R_{s1}R_{s2} + s\,\{\frac{R_{s2}}{C_1} + \frac{R_{s1}}{C_2}\} + 1/C_1C_2)}{s^2\,(\frac{R_1+R_{s1}}{R_1} + \frac{R_{s1}R_{s2}R_6}{R_2R_3R_5}) + s\,(\{\frac{R_{s2}}{C_1} + \frac{R_{s1}}{C_2}\}\frac{R_6}{R_2R_3R_5} + \frac{1}{R_1C_1}) + \frac{R_6}{C_1C_2R_2R_3R_5}}$$

The values of the switch resistances are chosen to meet two criteria: negligible pole frequency movement , and the two new zeroes introduced by the MOS switches must be as far outside the original filter bandwidth as possible to avoid disturbing the magnitude and phase response of the original filter in the passband. Assuming that all 3 MOS have identical on-resistance of 400Ω, the poles and zeroes of the modified circuit are: 2 poles at -90±j1000 radians (or pole frequency at 159 Hz), 2 real zeroes at -2.5E4 radians (much larger than pole frequency in radians). The frequency response is thus remarkably similar to the original response [21]. The gain peaking at 159 Hz is preserved, the roll-off outside the passband is identical up to 2 KHz (far outside the passband). The phase characteristics show more perturbation due to the zeroes when the frequency exceeds 1 KHz. Since this frequency is an order of magnitude larger than the filter bandwidth, the circuit response should not be affected significantly. This observation is more evident in the filter transient response to an input square wave of amplitude 1V and frequency 100 Hz. The two voltage outputs [21] are almost identical, showing that not only the phase perturbation is negligible but also the total harmonic distortion (THD) introduced by the non-linear MOS resistors is insignificant.

We have thus shown that the modified circuit performs almost identically to the original filter. The study of fault effects and the use of waveform scan techniques to observe these effects will be presented next. Two different faults are studied [21]:

1. A catastrophic fault in the second stage: C_2 has a broken connection to the opamp input. This is the type of catastrophic faults usually assumed in previous works , which includes open circuit, short circuit, and bridging faults. It is obvious that the circuit malfunctions with a square-wave output instead of the expected sine wave. The fault isolation procedure proceeds with 3 simulations: the filter in all-test mode (all switches set to disconnect capacitors from the circuit), stage 1 in test mode, and stage 2 in test mode. Note that stage 3, being a simple gain stage, is intentionally not modified in this case and does not have a corresponding test mode. The output waveform from the all-test mode, the output from testing stage 1, the output from testing stage 2 are used to confirm that stage 2 is faulty and to isolate to the capacitor C_2, which is the actual fault.

2. A non-catastrophic out-of-specification fault: R_5 has the incorrect value of $100K\Omega$ instead of the correct value $10K\Omega$. This fault is chosen to illustrate the application of the proposed DFT methodology in detecting and isolating faults due to parameter variations. The output waveform in normal mode, and the outputs from the all-test mode and from testing stage 1 confirm that the fault is thus likely to be with R_2 or R_5. Unfortunately, it becomes impossible at this point to isolate the fault further since both resistors always appear together in all expressions for the output waveforms, gains, and phases. The process stops at the identification of the fault (incorrect resistor value) and the possible faulty resistors (R_2 or R_5).

The specific advantages of this proposed methodology, as illustrated in the filter example, include:

(1) low overhead in terms of test devices.

(2) no AC noise coupling into the circuit since control signals are DC in either normal or test mode.

(3) well-defined DFT methodology suitable for CAD implementation at all levels (design, simulation, layout, test generation, and fault diagnosis).

(4) efficient fault diagnosis without full analog fault dictionary.

While the proposed DFT methodology was originally developed to address the problem of analog continuous-time active filters, the extensions to other circuits and other parametric procedures for fault isolation are possible. In SC circuits, the MOS devices are already an integral part of the SC structures, thus the application of this methodology will have lower overhead. The critical question here is not the extra devices as much as the timing methodology used during the test mode. A timing discipline based on the same idea of improving controllability and observability in SC filters has been studied and is presented in [36]. Extensions to SI and subthreshold analog circuits are reasonably straightforward: the timing methodology used for SC circuits is directly applicable to SI, and the DFT methodology for continuous-time circuits is applicable to subthreshold designs with modifications required to accommodate subthreshold scanning.

Analog testability and design-for-test are currently a subject of intense research due to the explosive advances in telecommunication and high-speed signal processors, where analog / digital hybrids constitute the fundamental design blocks. Works at the University of Washington, McGill University, University of Texas at Austin, MCC, IBM, and Bell Laboratories have shown early promises even though a comprehensive testable design methodology is still elusive. Efforts to establish test standards in conjunction with design techniques have also been initiated by the IEEE (e.g. P1149.4 Proposed Mixed-Signal Test Bus) to avoid the mistakes made in digital VLSI where design technologies far outstrip test technologies and result in a delay in product designs and manufacturing.

V. SUMMARY

This chapter has presented a succinct review of the state-of-the-art in the design of analog VLSI signal processors. The four major circuit techniques are explained and a discussion of design tools shows that it is still premature to speak of a completely analog VLSI system until the issues of tools for synthesis, simulation, and test are resolved. Current advances in technologies – e.g. BiCMOS, submicron CMOS, submicron bipolar, GaAs – are expected to impact circuit performance to benefit fundamental design disciplines such as switched-current and subthreshold techniques. In the short term, while the problems in analog system integration are dealt with in research laboratories, the design of high-performance hybrid analog / digital systems has accelerated and the expected progress in this area will propel analog VLSI into the spotlight as a strong candidate in future designs of signal processors.

206

REFERENCES

[1]   R. Hecht-Nielsen. *Neurocomputing* . Addison-Wesley, 1990.

[2]   C.A. Mead. *Analog VLSI and Neural Systems*. Addison-Wesley, 1989.

[3]   J. Harris *et al.* "Resistive fuses: Analog hardware for detecting discontinuities in early vision." In *Analog VLSI Implementation of Neural Systems*, eds. C. Mead and M. Ismail. Kluwers Academic Publishers, 1989.

[4]   C. Jutten, J. Herault, and A. Guerin. "IN.C.A.: an INdependent Component Analyzer based on an adaptive neuromimetic network." In *Artificial Intelligence and Cognitive Sciences*, eds. J. Demongeot *et al.* Manchester Press, 1988.

[5]   S. Bibyk and H. Ismail. "Issues in analog VLSI and MOS techniques for neural computing." In *Analog VLSI Implementation of Neural Systems*, eds. C. Mead and M. Ismail. Kluwers Academic Publishers, 1989.

[6]   P. Mueller *et al,* "Programmable analog neural computer and simulator." *Proc. IEEE Conf. on Advances in Neural Information Processing Systems*, Denver, 1989.

[7]   M.A. Sivilotti, M.A. Mahoward, and C.A. Mead. "Real-time visual computations using analog CMOS processing arrays." In *Advanced Research in VLSI – Proc. 1987 Stanford Conf.*, ed. P. Losleben. MIT Press, 1987.

[8]   P.R. Gray and R.G. Meyer. *Analysis and Design of Analog Integrated Circuits*. McGraw-Hill, 2nd ed., 1984.

[9]   R. Unbehauen and A. Cichocki. *MOS Switched-Capacitor and Continuous-Time Integrated Circuits and Systems*. Springer Verlag, 1989.

[10]   P.E. Allen and E. Sanchez-Sinencio. *Switched-Capacitor Circuits*. Van Nostrand Reinhold Co., 1984.

[11]   P.R. Gray, D.A. Hodges, and R.W. Brodersen, eds. *Analog MOS Integrated Circuits*. IEEE Press, 1980.

[12]   A. Fettweis, "Realization of general network functions using the resonant-transfer principle," *Proc. Fourth Asylomar Conf. Circuits & Systems,* pp. 663-666, Pacific Grove, CA.

[13]   D.L. Fried, "Analog sample-data filters," *IEEE J. Solid-State Circuits,* vol. SC-7, no. 4, pp. 302-304, August 1972.

[14]   J.B. Hughes, N.C. Bird, and I.C. Macbeth, "Switched currents – A new technique for analog sampled-data signal processing," *Proc. Int. Symp. Circuits & Systems*, pp. 1584-1587, 1989.

[15]   D.J. Allstot, T.S. Fiez, and G. Liang, "Design considerations for CMOS switched-current filters," *Proc. IEEE Custom Integrated Circuits Conf.*, pp. 8.1.1-8.1.4, 1990.

[16]   H.C. Yang, T.S. Fiez, and D.J. Allstot, "Current-feedthrough effects and cancellation techniques in switched-current circuits," *Proc. Int. Symp. Circuits & Systems*, pp. 3186-3188, 1990.

[17]   D.G. Nairn and C.A.T. Salama, "Current-mode algorithmic analog-to-digital converters," *IEEE J. Solid-State Circuits,* vol. 25, no. 4, pp. 997-1004, August 1990.

[18]   D.G. Nairn and C.A.T. Salama, "A ratio-independent algorithmic analog-to-digital converter combining current mode and dynamic techniques," *IEEE Trans. Circuits & Systems* , vol. 37, no. 3, pp. 319-325, March 1990.

[19]   R. Harjani, R.A. Rutenbar, and L.R. Carley, "OASYS: A framework for analog circuit synthesis," *IEEE Trans. CAD*, vol. 8, no. 12, pp. 1247-1266, December 1989.

[20]   A.T. Yang and J.T. Yao, "MISIM: A model-independent simulation environment, " *Proc. 7th Int. Conf. on the Numerical Analysis of Semiconductor Devices and Integrated Circuits*, 1991 (to be presented).

[21]   M. Soma, "A design-for-test methodology for active analog filters, " *Proc. IEEE Int. Test Conf.*, pp. 183-192, September 1990, Washington DC.

[22]   L. Rapisarda and R. DeCarlo, "Analog multifrequency fault diagnosis, " *IEEE Trans. Circuits Syst.,* vol. CAS-30, pp. 223-234, April 1983.

[23]   P.M. Lin and Y.S. Elcherif, "Analogue circuits fault dictionary - New approaches and implementation, " *Circuit Theory and Applications*, vol. 12, pp. 149-172, John Wiley & Sons, 1985.

[24]   S.L. Hakimi and K. Nakajima, "On a theory of t -fault diagnosable analog systems," *IEEE Trans. Circuits Syst.,* vol. CAS-31, pp. 946-951, November 1984.

[25]   Y. Togawa, T. Matsumoto, and H. Arai, "The T_F - equivalence class approach to analog fault diagnosis problems," *IEEE Trans. Circuits Syst.,* vol. CAS-33, pp. 992-1009, October 1986.

[26]   Z.F. Huang and R. Liu, "Analog fault diagnosis with tolerance," *IEEE Trans. Circuits Syst.,* vol. CAS-31, pp. 1332-1336, October 1986.

[27]   V. Visvanathan and A. Sangiovanni-Vincentelli, "Diagnosability on nonlinear circuits and systems -- Part I: The DC case," *IEEE Trans. Circuits Syst.,* vol. CAS-28, pp. 1093-1102, November 1981.

[28]   V. Visvanathan and A. Sangiovanni-Vincentelli, "A computational approach for the diagnosability of dynamical circuits," *IEEE Trans. CAD of Integrated Circuits Syst.,* vol. CAD-3, pp. 165-171, July 1984.

[29]   T.N. Trick and Y. Li, "A sensitivity based algorithm for fault isolation in analog circuits," *Proc. IEEE Intl. Symp. Circuits Syst.,* vol. 3, pp. 1098-1101, May 2-4, 1983.

[30]   A.E. Salama, J.A. Starzyk, and J.W. Bandler, "A unified decomposition approach for fault location in large analog circuits," *IEEE Trans. Circuits Syst.,* vol. CAS-31, pp. 609-622, July 1984.

[31]   C.L. Wey and R. Saeks, "On the implementation of an analog ATPG: The linear case," *IEEE Trans. Instrum. Meas.,* vol. IM-34, pp. 442-449, September 1985.

[32]   Z.F. Huang, C.S. Lin, and R. Liu, "Node-fault diagnosis and a design of testability," *IEEE Trans. Circuits Syst.,* vol. CAS-30, pp. 257-265, May 1983.

[33]   G.J. Hemink, B.W. Meijer, and H.G. Kerkhoff, "TASTE: A tool for analog system testability evaluation," *Proc. IEEE Intl. Test Conf.,* pp. 829-839, September 12-14, 1988, Washington DC.

[34] M.J. Marlett and J.A. Abraham, "DC_IATP: An iterative analog circuit test generation program for generating DC single pattern tests," *Proc. IEEE Intl. Test Conf.*, pp. 839-845, September 12-14, 1988, Washington DC.

[35] K.D. Wagner and T.W. Williams, "Design for testability of mixed signal integrated circuits," *Proc. IEEE Intl. Test Conf.*, pp. 823-829, September 12-14, 1988, Washington DC.

[36] M. Soma and V. Kolarik, "A design-for-test technique for switched-capacitor circuits," *Proc. IEEE VLSI Test Symp.*, Cherry Hill, NJ, 1994.

8

SWITCHED-CAPACITOR PARALLEL DISTRIBUTED PROCESSING NETWORK FOR SPEECH RECOGNITION

Yoshihiko Horio*, Hiroyuki Takase** and Shogo Nakamura*

*Department of Electronic Engineering, Tokyo Denki University 2–2, Kanda-Nishiki-cho, Chiyoda-ku, Tokyo, 101 Japan. **Computer Works, Mitsubishi Electric Co. 325, Kamimachiya, Kamakura, Kanagawa, 247, Japan.

1. INTRODUCTION

Parallel distributed processing (PDP) systems such as neural networks and associative networks are currently receiving a great deal of interest within a variety of areas [1]. Analog signal processing, especially using VLSI circuit techniques, is suitable for implementation of such systems because of the real-time and parallel processing capability inherent in analog circuitry [2–6, 18].

One of the great application areas of the analog PDP system is speech processing. In this chapter, a new class of Switched-Capacitor (SC) PDP network system is proposed in an attempt to achieve a real-time monosyllabic speech recognition without using the reference pattern dictionary [7–9, 19, 20]. This system contains the idea of neural network architecture: for instance, a massively parallel processing, variable connection weights, a learning property, and so on. Furthermore, because the recognition of the speech signal can be utilized in the network without any timer circuit.

In a speech recognition system, it is very important to choose and extract a phonetic feature. In particular, a real-time extraction is desirable in an analog PDP system. Many studies on the feature extraction of the speech signal have been done, and several good features have recently become available [10, 11]. In particular, the shape of the spectrum and its time transition pattern are very important features. Moreover, the number, location, and time-ordered variation of peaks of the spectrum (local peak transition pattern) include important information and express unique and specific features of each word. Fortunately, this local peak transition pattern can be extracted in real-time using SC circuits as

210

will be mentioned in Section 2. As a consequence, the local peak transition pattern of the spectrum is used as the phonetic information in the following system.

The overall block diagram of the speech recognition PDP network system is shown in Figure 1. As shown in the figure, the system consists of a preprocessor, recognition networks and a post-processor. The SC preprocessor emphasizes the local peaks of the input speech spectrum and extracts the local peak transition pattern, as reviewed shortly in Section 2. Each network in Figure 1 recognizes one word which corresponds to that network, and the input speech is applied to all of the networks in parallel. Then each network automatically tracks the

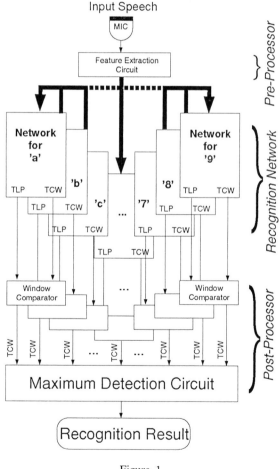

Figure 1

local peak transition pattern and computes two matching scores, Total

Connection Weight (TCW) and Total Location Point (TLP). The basic recognition algorithm and automatic tracking method of the local peak transition pattern using the SC networks are described in Section 3. These two recognition parameters, TCW and TLP, which are introduced to accomplish the recognition without using an additional matching process are detailed in Section 4, with the post-processor which determines the final recognition result using these parameters. In order to make the network suitable for the analog VLSI implementation and to improve the recognition performance, the networks are pre-tuned as described in Section 5. The weight learning rule is given in Section 6, and the circuit implementation of the PDP system is summarized in Section 7. Finally, in Section 8, simulation and experimental results are reported.

2. PREPROCESSOR

The block diagram of the preprocessor is shown in Figure 2. As shown in the figure, the first building component is a conventional AGC circuit by which the input speech signal is level-adjusted, and a simple 1st-order high-pass filter is used as a emphasis filter. Furthermore, the input signal is band-limited below 10 kHz by an anti-aliasing filter.

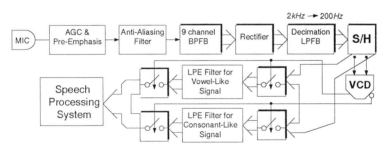

Figure 2

After this high-pass filtering, a SC 9ch band-pass filter bank extracts the input speech spectrum. Each band-pass filter is a SC 12th-order LDI filter which is designed by using FILTOR2 [12] and SWITCAP [13] programs. The overall response of this filter bank is shown in Figure 3. As shown in the figure, the response is different from a conventional speech processing band-pass filter bank in which each filter response is equally spaced in log scale. In contrast, each response in Figure 3 is specially estimated by the simulations with over 50 speech data of 5 people in order to extract the local peak pattern better than the conventional filter banks.

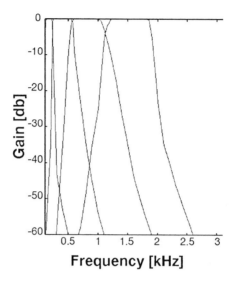

Figure 3

The fourth component of the preprocessor is a 9ch offset-cancelling SC full-wave rectifier [14] and the fifth one is a 9ch SC CIC decimation low-pass filter bank. The block diagram of the CIC decimation filter [15] and its SC building blocks are shown in Figure 4 and 5, respectively. By this decimation filter bank, the sampling frequency is lowered from 20 kHz to 200 Hz, and this decimation low-pass filter bank also acts as an anti-aliasing filter bank for the following stage.

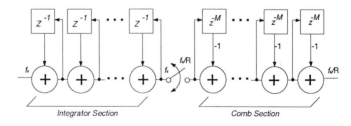

Figure 4

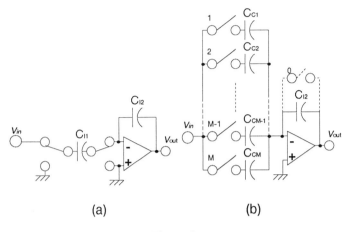

(a) (b)

Figure 5

After the sampled-and-hold (S/H) Section, the local peaks of the spectrum are extracted and emphasized by the Local Peak Extraction (LPE) filter which consists of linear phase CIC filters as shown in Figure 6. As shown in Figure 2, the preprocessor has two different LPE filters. One of these is designed for vowel-like signal and another is for consonant-like signal, and they are switched so that one of these LPE filters is used at a time. The selection (switching) of the LPE filters is determined by using the SC Vowel-Consonant Detector (VCD) [8].

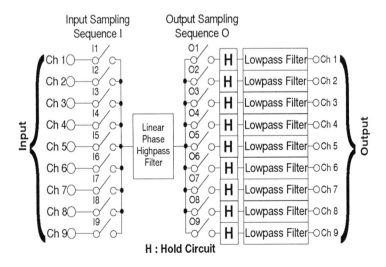

Figure 6

Finally, the local peak transition pattern of the input speech spectrum is obtained at the output of the LPE filter. This output is applied to the following speech processing network in parallel. For more detail of the preprocessor, see Horio et al [8,17].

3. NETWORK ARCHITECTURE

3.1. Basic Network Architecture

The key concept of our recognition algorithm is the automatic tracking of the local peak transition pattern by using the SC Parallel Distributed Processing (PDP) networks.

The recognition network consists of unit cells in rows and columns as shown in Figure 7. Each cell, $CE_{i,j}$, has its own 9–dimensional *State Vector*, $V_{i,j}$ as

$$V_{i,j} = \left(v_1^{i,j}, v_2^{1,j}, ..., v_l^{i,j}, ..., v_9^{i,j} \right), \tag{1}$$

where v_l = +1 or –1, l is a channel number, i is a row number and j is a column number in the network. Cells within the same row have different state-vectors and each cell is connected to those in adjacent rows. Then umber of cells contained in the ith row, m_i, is optimized by pre-tuning as described in Section 5.

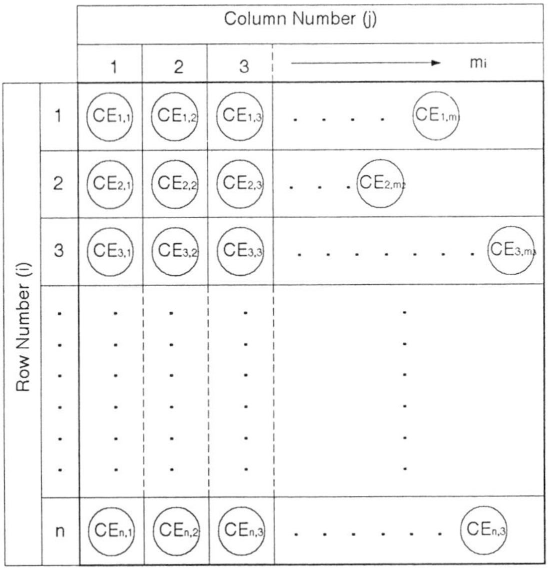

Figure 7

3.2. Tracking of Local Peak Pattern

Because the input signal to each cell is the output of the 9ch preprocessor, it can be also expressed as a 9–dimensional *Input Vector*, S_k, as

$$S_k = \left(s_1^k, s_2^k, ..., s_l^k, ..., s_9^k \right), \qquad (2)$$

where s_l has a continuous value, l is a channel number, and k stands for the kth sample of the input sequence. In each cell, the inner product of S_k and $V_{i,j}$ is calculated as

$$IP_{i,j,k} = \left(S_k, V_{i,j} \right) = \sum_{n=1}^{9} s_n^k * v_n^{i,j}, \qquad (3)$$

Let us assume the following conditions within a network.

Only one cell can be **fired** (activated) at a time, and the cell, $CE_{p,q}$ ($1 \leq p < n$, $1 \leq q < m_p$), is now firing.

The active cell, $CE_{p,q}$ and all cells in the next row, $CE_{p+1,j}$ for all j, calculate the inner products.

$CE_{p,q}$ transfers its inner product value, $IP_{p,q,k}$, to all cells in the next row.

Under these conditions, the tracking of the local peak transition pattern is accomplished as follows. Each cell in the $p+1$th row compares its own inner product value, $IP_{p+1,j,k}$, to that of the firing cell, $IP_{p,q,k}$. If $IP_{p+1,r,k}$ ($1 \le r < m_{p+1} = \text{MAX}[IP_{p+1,j,k}]$ and $IP_{p+1,r,k} > 1.1 IP_{p,q,k}$, then $\mathbf{CE}_{p+1,r}$ will fire next, and at the same time, $\mathbf{CE}_{p,q}$ will quit firing. Otherwise, cell $\mathbf{CE}_{p,q}$ continues firing. This process is repeated until the end of the input signal. As a result, a *state vector transition pattern* (SVTP) is obtained as a locus of the state vector, $V_{i,j}$, of the cells which have been fired during the tracking process.

As a consequence, the local peak transition pattern of the input speech has been tracked, because it relates to the SVTP very closely. In other words, by utilizing SVTP, the input speech can be recognized. However, the problem is how to identify SVTP without using any additional matching process, which is hard to implemented as the analog circuitry. A solution, post-processing, is given in the next Section.

4. POST-PROCESSOR

4.1. Recognition Parameters

A general method of speech recognition, that is, matching, is dependent on how the distance between the input word data and the dictionary word data is calculated. However, these calculations may be too complicated for analog systems, because the conventional matching methods are algorithmic, and further, they need a huge space of memory. To avoid the additional matching process, two convenient parameters are proposed.

The first parameter is the *total location point* (TLP). To calculate TLP, the *cell location point*, $lp_{i,j}$, which has a fixed value related to the location of the cell in a network, is assigned to each cell. Namely, $lp_{i,j}$ is the function of row number i and column number j. Furthermore, in the tracking process, if cell $\mathbf{CE}_{i,j}$ is activated during the J clock cycles, the final location point. $LP_{i,j}$, of $\mathbf{CE}_{i,j}$ is calculated as

$$LP_{i,j} = lp_{i,j}(1 + k * J), \tag{4}$$

where k is a constant. To obtain the TLP at output node of the network, the $LP_{i,j}$ of all fixed cells are summed as

$$TLP = \sum_{i=1}^{end} (LP_{i,j} \, of \, the \, fired \, cells), \qquad (5)$$

The *end* in the above equation expresses the row number at which the tracking ends. For instance, if the input signal closely matches with the characteristic of one of the networks, the tracking in this network will end near the bottom of the network, that is, $end \approx n$.

The second parameter is the *total connection weight* (TCW). TCW is simply the total sum of the weight values of connections which have been selected during the tracking process. Each cell adds the connection weights of the preceding fired cells to its own, and passes the sum to the next cell. Finally, the total sum in a network is obtained as

$$TCW = \sum_{i=1}^{end} (Weight \, of \, Used \, Connection), \qquad (6)$$

where *end* is the same row number in (5).

4.2. Post-Processor

The recognition result is obtained by two post-processors using TLP and TCW data. The first post-processor (PostP1) consists of a window comparator and a selective switch as shown in Figure 8. The location points in each network are pre-tuned using statistical data so that the TLP of each network is classified into one of the ten network groups. Each group contains networks for easily distinguishable words in it, whereas the networks for hardly distinguishable ones, for instance, networks for *d*, *b*, *t* and *p* or those for *m* and *n*, belong to the different network groups respectively. On the other hand, the window comparators are also pre-tuned in order to classify the resulting TLP into one of the ten groups which correspond to these ten network groups. Finally, under these conditions, the PostP1 selects one of the ten network groups by using the TLP data, and as a consequence, the recognition result is roughly determined.

The second processor (PostP2) is a simple maximum detection circuit using the winner-take-all circuit in [16], and determines the final recognition result by choosing the most plausible network using the TCW data.

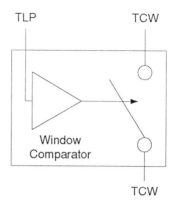

TLP   TCW

Window
Comparator

TCW

Figure 8

5. PRE-TUNING OF THE NETWORK

The current target of the proposed PDP system is to recognize thirty-six utterances including the twenty-six letters of the alphabet and ten digits. For this purpose, many cells could be omitted, making the whole system compact. In addition, each network can be tuned to maximize the recognition rate.

First, the cell reduction process is presented. The original networks have 512 cells in each row; however, not all of them are used in the recognition process. After many simulations using the original networks for each word, a histogram of the fired cells, that is, the *fire rate*, is obtained as in Figure 9(a). Cells never or rarely fired in the simulations are removed from the network preserving the recognition performance. As a result, the average number of cells per row is reduced to twenty.

In the resulting compact networks, the cells are allocated as follows. Using the fire rate of each cell mentioned above, the most frequently activated cell is placed at the center of the column in each row; the second most activated cell is placed next to the center, the third one is placed on the other side of the center, and so on, as in Figure 9(b).

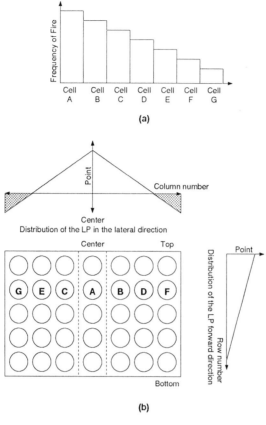

Figure 9

Moreover, the location points in row direction (vertical direction) are allotted according to a monotonically decreasing function, and those within each column (lateral direction) have triangle shape distribution function as shown in Figure 9(b). Furthermore, the triangle shape distribution of the point creates the inhibiting connections in the negative region of points as shown by the dark region in the figure. The cells which are sensitive to the hardly distinguishable utterances from the objective word of each network are allocated in this inhibition region so as to prevent misrecognition. For example, in the network for *b*, the cells which are sensitive to the similar utterances to *b*, such as *d*, *p* and *t*, are allocated in the inhibition region.

In addition to these location point arrangements, the location point is pre-tuned for PostP1 as described in the above Section. As a result, the resulting TLP after recognition phase will take the value close to a certain fixed (pre-tuned) point for a correct (target) utterance and far from that for incorrect ones.

6. WEIGHT LEARNING

The connection weights are learned by the improved Hebb's law during the training phase. The weights of selected connecting paths are increased by one small step, $+\Delta w$, while weights of unused connection paths are decreased by $-\Delta w$. However, all weights have upper and lower limits. The learning will be able to be done during not only in the training phase but also in the recognition phase if a certain feedback circuitry is added to the network from the post-processor.

7. SC IMPLEMENTATION OF THE NETWORK

The cell in the recognition network is SC implemented as shown in Figure 10. In the figure, the upper part containing I_1, I_2 and O_1, is the weight circuit to calculate TCW. The PCA in the figure stands for a programmable capacitor array, and the learning circuit consists of up/down counters as shown in Figure 11.

The middle part in Figure 10 with I_{2-5} is the cell state decision circuit. The inner product (IP) circuit which contains the state vector, $V_{i,j}$, of the cell is depicted in Figure 12. The maximum detection circuit in the figure is a part of the winner-take-all circuit in [16].

Finally, the lower part of Figure 10 with I_6 and O_6 is the location point circuit which is detailed in Figure 13.

Using the unit cells shown in Figure 10, the whole recognition network is constructed as shown in Figure 14. In the figure, I_c is a current source for the winner-take-all circuit.

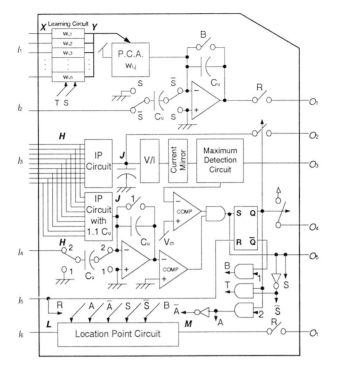

Figure 10

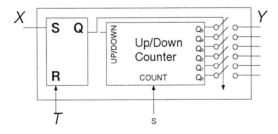

Figure 11

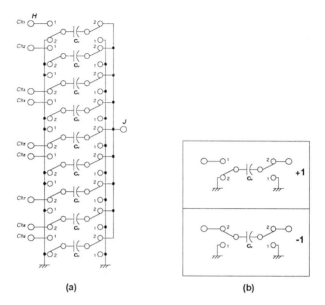

(a)   (b)

Figure 12

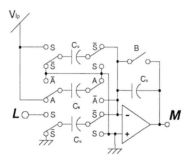

Figure 13

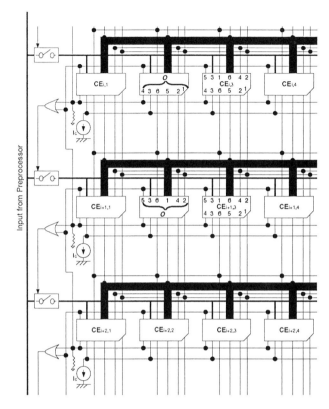

Figure 14

8. SIMULATION AND EXPERIMENTAL RESULTS

Although the goal of this system is speaker-independent word recognition, our system is tuned for a specific male speaker at present. Data for ten digits and the twenty-six letters of the alphabet were used thirty times for training and then twenty times for recognition. The learning step $\Delta w = \pm 0.5$, and upper and lower weight limits of each connection weight are ± 20. The pre-tuned network has less than forty rows and thirty columns. As a result, the present average recognition rate is about 88% for all ten digits and the twenty-six letters of the alphabet. The recognition rate improvement according to learning is shown in Figure 15.

224

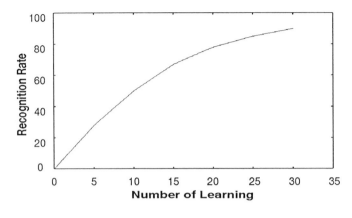

Figure 15

The SC preprocessor and post-processor have been implemented with discrete elements. In addition, the SC has been implemented also using the discrete elements, and its operating characteristics have been confirmed. Furthermore, a small SC PDP system which recognizes some digits has been implemented using the discrete elements and successfully tested. The whole system will be soon available, so the systematic experimental results including the statistical recognition rate will be reported soon.

9. CONCLUSIONS

A real-time speech recognition system using the SC PDP network has been proposed. The preprocessor which extracts the local peak pattern of the input speech has been proposed; and a method for tracking the local peak transition patterns using the SC networks has been described. Moreover, in order to recognize the input speech without an external matching system, a network tuning procedure and the post-processor have been proposed.

The recognition rate as a function of the parameter k in (4) is shown in Figure 16. As shown in the figure, as a future problem, the optimization of the parameter k in (4) should be done. Moreover, further number of learning should be done to obtain the final recognition rate.

Acknowledgement

This research was partly supported by the Research Institute for Technology, Tokyo Denki University, under Project Q63–S10. The authors also would like to express their gratitude to the Foundation of Ando Laboratory for their encouragement and financial support.

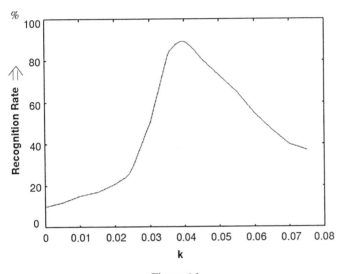

Figure 16

Reference

[1] R. Lippmann, "An Introduction to Computing with Neural Nets", IEEE Magazine, April 1987, pp. 4–22.

[2] J. Hopfield et al, "Computing with Neural Circuits: A Model", Science, Vol. 233, August 8, 1986, pp. 625–633.

[3] D. Tank et al, "Simple Neural Optimization Networks: An A/D Converter, Signal Decision Circuit, and Linear Programming Circuit", IEEE Trans. on CAS., Vol. 33, No. 5, May 1986, pp. 533–541.

[4] Y. Tsividis et al., "Analogue Circuits for Variable-Synapse Electronic Neural Networks", Electronics Letters, Vol. 23, No. 24, Nov. 1987, pp. 1313–1314.

[5] P.K. Houselander et al., "Current Mode Analogue Circuit for Implementing Artificial Neural Networks", Electronics Letters, Vol. 24, No. 10, May 1988, pp. 630–631.

[6] P.K. Houselander et al., "Current Mode Analogue Circuit for Implementing Artificial Neural Networks", Electronics Letters, Vol. 24, No. 10, May 1988, pp. 630–631.

[7] Y. Horio, et al., "Speech Recognition Network with SC Neuron-Like Components", IEEE Proc. of ISCAS'88, 1988, pp. 495–498.

[8] Y. Horio, et al., "Switched-Capacitor PreProcessor for Speech Processing Using SC CIC Filter", IEEE Proc. of ISCAS'89, 1989, pp. 1311–1314.

[9] H. Takase et al., "Speech Recognition Algorithm Based on a Switched Capacitor Network System", IEEE Proc. of ECCTD'89, 1989, pp. 492–496.

[10] T. Parsons, "Voice and Speech Processing", McGraw-Hill Book Company, 1986.

[11] J.N. Holmes, "Speech Synthesis and Recognition", Van Nostrand Reinhold, 1988.

[12] W.M. Snelgrove et al., "FILTOR2 User's Manual", University of Toronto Press, Toronto, Canada, 1978.

[13] S.C. Fang, et al., "SWITCAP: A Switched-Capacitor Network Analysis Program, Part I: Basic Features", IEEE Circuit & System Magazine, September 1993, pp.10–40.

[14] R. Gregorian et al., Analog Integrated Circuits for Signal Processing, John Wiley & Sons, New York, 1986.

[15] E.B. Hogenauer, "An Economical Class of Digital Filters for Decimation and Interpolation", IEEE Trans. on ASSP, Vol. ASSP-29, No. 2, April 1981, pp. 155–162.

[16] J. Lazarro et al., "Winner-Take-All Networks of O(N) Complexity", Advances in Neural Information Processing Systems 1, Morgan Kaufmann Publishers, San Mateo, CA, 1989, pp. 703–711.

[17] Y. Horio et al., "A Switched-Capacitor PreProcessor For Speech Recognition", Analog Integrated Circuits & Signal Processing, an International Journal, Vol. 2, No. 2, pp. 79-94, Jan. 1992.

[18] Artificial Neural Networks, Edgar Sanchez-Sinencio and Clifford Lau eds., IEEE Press Book, 1992.

[19] H. Sugawara et al, "A New Pre-Processing Filter for a Network Based Speech Recognition", Proc. of 5th European Signal Processing Conference (EUSIPCO), pp. 1247-1250, Sept. 1990.

[20] Youichi Kohno et al, "Speech Recognition Using State Networks", PRoc. of the Internatinal Conference on Signal Processing Applications and Technology, pp. 1052-1055, Nov. 1992.

Index